P

FOR POLLUTION

YOUR GUIDE TO POLLUTION AND HOW TO STOP IT

Brian Price

GREEN
PRINT

First published in 1991 by
Green Print
an imprint of The Merlin Press
10 Malden Road, London NW5 3HR

© Brian Price

ISBN 1 85425 059 0

1 2 3 4 5 6 7 8 9 10 : : 99 98 97 96 95 94 93 92 91

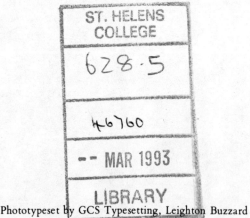
Phototypeset by GCS Typesetting, Leighton Buzzard
Printed in England by Biddles Ltd., Guildford, Surrey
on recycled paper

P For Pollution

Brian Price is a freelance pollution consultant and writer. He lectures on the Post-Graduate Diploma in Environmental Health at Bristol Polytechnic and is an Open University tutor. He writes and broadcasts frequently on pollution topics and recent works include *C for Chemicals* (with Mike Birkin) and *Hazardous Waste* – a report for the Financial Times Business Information Unit. He is a chemist and lives in Weston-super-Mare with his wife and children.

PREFACE

This book is an attempt to provide the concerned reader with a simple guide to the major pollution problems of today. I have attempted to steer clear of excessive jargon and to keep the science simple – Chapter 1 is an outline of the science involved in pollution but the book should make sense without it. I have also attempted to present controversies objectively although this is sometimes difficult – politics are as much a part of pollution as is chemistry. Any comments on my success, or otherwise, in achieving these aims will be welcome, c/o the publishers.

I would like to record my thanks to my wife, Jen, for her extremely helpful comments on the manuscript. I would also like to thank Jon Carpenter at Green Print for his patience.

Brian Price   *December 1990*

To Simon, Rohan and Caitlin
May they inherit a cleaner, safer world.

CONTENTS

INTRODUCTION

Pollution is one of the major problems facing humanity today. Scarcely a week goes by without a new pollution crisis hitting the headlines. Global warming, the ozone layer, toxic waste and polluted drinking water have all featured on the political agenda in recent years and environmental groups have experienced a tremendous upsurge in support.

But what are the real problems behind the rhetoric? Are we really threatening the future of the planet with our wastes? This book attempts to answer these questions and to provide a factual background which will enable readers to make up their own minds.

Before addressing the problems themselves it is useful to establish what we mean by 'pollution'. There are several definitions in use but a convenient one is *the introduction of something into the environment, by human activities, which causes harm.* This is a very general definition since it includes heat and noise as well as chemicals and sewage, the type of material which we normally identify with pollution. The definition does not just apply to synthetic materials: natural substances discharged in excessive quantities can do just as much, if not more, harm than many artificial ones. Furthermore, human activities have to be involved if pollution is to be confirmed: natural processes are exempt from this definition. Finally, a crucial characteristic of pollution is that harm of some kind must be caused, be it human illness, the death of wildlife or the corrosion of a cathedral. The mere presence of a substance in the environment may be contamination but unless it is harmful, it is not pollution.

Armed with a working definition of pollution we can now consider its nature, behaviour and effects. Chapter 1 is an outline of science as it applies to pollution. It is followed by a chapter on pathways in the environment. Chapter 3 considers the effects of pollutants. Part 2 is a directory of pollutants and processes. In Part 3, a chapter on the law is followed by suggestions as to how individuals can help. A list of useful organisations and a bibliography make up the appendices. There may well be some duplication of examples and data; for this I apologise, but I want to ensure that the casual reader can find the required information when simply dipping into the book.

PART 1

POLLUTION AND HOW IT WORKS

1: SCIENCE AND POLLUTION

You do not need to be a highly qualified scientist to understand how and why pollution causes problems, but it is useful to know some basic science. This is not as alarming a prospect as it may seem since many of the ideas involved are reasonably straightforward – experts sometimes hide behind a smokescreen of jargon and complexity which is not really necessary. This chapter is designed to introduce some basic chemistry, physics and biology: it is not intended to be a foundation course in science but it should help the reader to follow some of the scientific arguments in the pollution sphere. If the science still seems difficult it does not matter too much as the rest of the book attempts to avoid the excessive use of jargon.

CHEMISTRY

Elements and atoms

One of the basic ideas of chemistry is the atomic theory. The essential aspect of this is that all the materials around us, from air to granite, are made up of **atoms**. These are minute particles which, until the early part of this century, were thought to be indivisible – indeed the word atom comes from a Greek word meaning un-cuttable. We now know that atoms can be split and this is the province of the nuclear physicist; for the purposes of chemistry they remain indivisible. There are about 107 different types of atoms, some fifteen of which are synthetic, made in nuclear reactors and similar devices. The remaining 92 are naturally occurring. Each type of atom is characteristic of a particular **element** and every substance is composed of various combinations of elements. Each element has well-defined chemical properties – for example it may be a metal, it may combine readily with certain other elements or it may be very reluctant to combine with anything at all. Examples of common elements are aluminium, oxygen, carbon, iron and sulphur. All atoms of the same element behave the same way in chemical processes but some may weigh slightly more or less than others – when this occurs the element is said to have isotopes.

Elements cannot normally be created or destroyed. A consequence of this is that pollutants which are also elements, such as lead and mercury, are never broken down in the environment: they may combine with varying proportions of other elements as environmental conditions change, but they retain their elemental identity and often their toxic properties. The only way in which one element can be converted into another is by radioactive decay (see p. 103) or in a nuclear reactor.

Compounds and molecules

Most atoms have a tendency to combine with other atoms and they do this in fixed proportions to form **compounds**. Thus a compound has a fixed composition – the material hydrogen chloride, for instance, always consists of equal numbers of hydrogen and chlorine atoms. In many cases the atoms of a compound's constituent elements join up to form **molecules** – the smallest possible particle of the compound. The simplest molecules consist of two identical atoms – in other words some elements exist as molecules rather than uncombined atoms. The hydrogen molecule consists of two hydrogen atoms while the normal form of oxygen consists of two-atom molecules. Ozone, a special form of oxygen, consists of molecules made up of three atoms.

Some compounds – known as ionic compounds – do not consist of molecules but are made up of **ion** pairs or groups of ions. An ion is an atom which has gained or lost an electric charge and an ion pair consists of two oppositely charged ions. (Electric charge is a difficult concept to grasp but it is familiar as the 'static electricity' which builds up when some materials are rubbed together, such as the soles of some shoes against certain types of carpet. Electric charge may be positive or negative in nature). The charges on the ions attract each other and balance out, so that for every positive charge there is a negative. In some instances two negative charges may be needed to balance an ion with a double positive charge, or vice versa. This type of compound usually forms crystals: common salt, for instance, forms cube-shaped crystals consisting of a pattern of alternating sodium and chloride ions.

Solutions

A solution is formed when a solid or gas dissolves in a liquid – the **solvent** – and the commonest solvent is water. The degree to which a substance dissolves in water (its solubility) is of great importance in environmental terms since it can determine what happens to it when it is released. A highly soluble material such as nitrate, for instance, is easily washed from fields into waterways by rainfall. Metal compounds called hydroxides are usually much less soluble so metals can be removed from effluents, to a large extent, by converting them to hydroxides and allowing them to settle out (**precipitate**).

The solubility of a substance in oils and fats is also an important property since this can determine whether or not it is stored in the bodies of animals which have extensive fat deposits. A material such as DDT, which is poorly soluble in water but readily soluble in fat, will be concentrated from the water by fish and stored in fat at much higher concentrations than were present in the water.

When ionic compounds dissolve in water the ion pairs or groups of ions separate. This, incidentally, enables the solution to carry an electric current. Some substances produce large numbers of hydrogen ions when they dissolve: these are known as **acids**. Others, known as **alkalis**, produce large numbers of hydroxyl ions – a charged combination of one hydrogen and one oxygen atom. In pure water the number of hydrogen ions balances the number of hydroxyl ions and the liquid is neutral. The acidity of a solution is measured on the **pH scale** with acids having low pH values and alkalis having high values. The mid-point of the scale, neutrality, is pH 7.

As a general rule, fatty and oily materials do not dissolve in water but are soluble in **organic solvents** such as benzene or trichloroethane. These organic solvents do not mix with water and substances which dissolve in water do not usually dissolve in them. Some substances can bridge the gap between the two classes of material and these are familiar as soaps and detergents. A detergent molecule has two parts to it: a chain or group of atoms which dissolves in oils and an end which is ionic and therefore soluble in water. Thus if a mixture of oil and water is shaken with a detergent the oily ends of the detergent molecules will dissolve in the oil, the other ends will dissolve in water, and an emulsion or **suspension** (see below) of the oil in water can be formed. Thus grease stains can be removed from clothes and chip fat can be cleaned off plates.

When materials are in suspension they are present as particles much larger than ions or molecules and are prevented from precipitating out by the movement of the water – as in a river – or by other forces in the liquid. Particles of minerals are often carried long distances in suspension, as are some pollutants from sewage works. When the water slows down or the chemistry of the water changes the particles fall to the bottom.

Chemical reactions

A **chemical reaction** occurs when atoms and molecules combine or recombine to form different substances. For instance, hydrogen and oxygen molecules will combine together to form water molecules, sulphur atoms will combine with oxygen molecules to form sulphur dioxide, and sodium hydroxide will react with hydrochloric acid to form sodium chloride and water. Chemical reactions often occur because the end products are more stable than the mixture of starting materials. As they do so energy is released – the reactions listed above all involve the production of heat and the combination of hydrogen and oxygen can be explosive. Reactions which give out energy as they occur are said to be **exothermic** whereas those which involve the absorption of energy are **endothermic**. Most exothermic reactions need the input of a small amount of energy to get them started – a spark, for instance, will start the reaction between petrol vapour and oxygen in a car engine.

One of the most familiar exothermic reactions is **combustion**, the combination of carbon and hydrogen, in fuels such as gas and oil, with oxygen in the air. Coal is mainly carbon and produces carbon dioxide as its main combustion product, but oil and gas produce carbon dioxide and water since they also contain hydrogen. Most organic compounds will burn under appropriate conditions.

Combustion is an example of a type of reaction called **oxidation**. Reactions of this type often involve the combination of another material with oxygen. The opposite of oxidation is **reduction**, which often involves the removal of oxygen from a compound. Oxidation does not necessarily involve burning – for instance the element iron can exist in solution in an oxidised or a reduced form. Water accumulating in old mine workings contains iron in the reduced form but when it is discharged into a river the oxygen in the water converts it into the less soluble oxidised form and a red-brown precipitate is formed. This process deoxygenates the river and such waters are highly polluting. Respiration by living things is a form of oxidation and in most instances this involves the conversion of substances such as glucose – consisting of carbon, hydrogen and oxygen – to carbon dioxide and water by combining them with oxygen. The energy released is used for living processes such as growth, movement, heart beats etc.

Catalysts

Sometimes a chemical reaction may occur very slowly and it is desirable to speed it up. This is achieved in industry and in living things by the use of **catalysts**. A catalyst is a material which speeds up a reaction without itself being consumed – examples include platinum, used to clean up vehicle exhaust emissions by promoting reactions which convert harmful emissions to less hazardous gases, and the thousands of enzymes which catalyse biochemical processes in living organisms. Catalysts can only influence the rate of a reaction: they cannot cause something to happen which would not happen anyway over time.

Complex molecules

The element **carbon** is unique because its atoms can join together in long chains of almost unlimited length. Various other elements attach to these chains – hydrogen is the commonest – and millions of different combinations are possible. These molecules are the subject of **organic** chemistry and organic chemicals are the basis of life. Proteins, fats, carbohydrates and the molecules which make up the genetic code are all organic chemicals, as are plastics, textiles, most drugs and pesticides.

Because of carbon's unique properties there are many more organic compounds than there are **inorganic** substances – those based on elements other than carbon. It was once thought that organic chemicals could only be made by living things, but this idea was disproved when a simple organic compound, urea, was made completely artificially. Nowadays, the vast majority of organic compounds of industrial significance are made synthetically, usually from petroleum.

Some complex molecules consist of a repeating pattern of identical sub-units, in much the same way that a wall consists of a repeating pattern of bricks. Such substances are called **polymers** and a familiar example is the plastic PVC, or polyvinyl chloride. This consists of vast numbers of vinyl chloride units linked together in three dimensions to form an extended network. **Proteins** are also polymers although they do not consist of repeats of single units – there are around twenty different types of building blocks, called **amino-acids**, to choose from.

PHYSICS

Energy

A fundamental concept which links chemistry, biology and physics is **energy**. Energy enables things to happen as it is converted from one form to another and without it the universe would be dark, immobile and lifeless. Energy is used in many forms, from the food we eat to the output of a power station, and modern civilisation is dependent on the ready availability of abundant energy supplies. Energy can be stored in many ways too: as chemical compounds such as sugars or explosives, as potential energy when water is held behind an upland dam forming part of a hydroelectric power station, as motion in a flywheel, or as heat in a storage radiator.

The behaviour of energy is described by the **laws of thermo-dynamics**. Among these principles are two basic ideas: energy cannot be created or destroyed, and spontaneous energy flow is one-way. When we burn oil or use food to provide us with the energy to move, we are not creating energy, we are simply converting it from stored chemical energy to another form – heat or motion. Similarly, when we stop a car we are not destroying its energy of motion, we are converting it to heat in the brakes and tyres. The only way in which energy can actually be created is in nuclear reactions such as those taking place in atomic bombs. In these cases matter is converted to energy so the law should really state that energy and matter taken together cannot be created or destroyed.

The one-way spontaneous flow of energy can be summarised by the phrase 'Things cool down'. In other words, heat flows from a hot area to a cold area and the temperature drops unless more energy is supplied to maintain it. Moving things will slow down, because of friction and similar effects, unless energy is constantly supplied and living things will die without food. Some types of energy are regarded as being high-grade – e.g. high temperature heat, light and electricity. As they are used, their energy content becomes degraded and they all end up as low temperature heat. Organised systems become disorganised over time unless energy is constantly supplied, a principle which applies to everything from an ecosystem to a teenager's bedroom.

There are three main sources of energy for the planet: (1) radioactive decay in deep rocks, (2) energy from the sun, and (3) the energy of the tides which comes from the rotation in the earth/moon/sun system. Radioactive decay causes a flow of heat outwards through the earth's crust and also fuels geysers and other geothermal heating phenomena. The sun's rays warm up the surface and are used by plants, hence providing the basis of all our food. They power the wind, waves and rain and are also stored in fossil fuels such as coal and oil. The vast majority of human energy use depends on these sources, the only other type of current significance being nuclear power which exploits the energy stored inside the atoms of fuels such as uranium.

Energy considerations are vital in pollution issues for several reasons. Energy use nearly always involves a potential for pollution or other forms of environmental harm. No source is totally benign although some are much worse than others. Controlling pollution may well involve extra energy use – cars fitted with catalysts to reduce exhaust fumes burn more petrol and power stations fitted with devices to remove sulphur dioxide burn more coal per unit of electrical output. There are some exceptions – recycling oil which would otherwise be discarded or improving engine efficiencies to cut emissions, will both save energy – but dealing with a widely dispersed environmental contaminant (such as cleaning up a polluted water supply) will involve the use of much energy. Preventing the release of the material into the environment is thus far preferable to cleaning it up afterwards.

Electromagnetic radiation

One particular form of energy is **electromagnetic radiation**. This is a very broad class of phenomena, covering X-rays, infra-red (heat) radiation, visible light, ultraviolet, microwaves and radio waves among other things. Radiation transfers energy from one place to another without the need for an intervening conductor such as an electric wire and this is clearly the way in which we receive energy from the sun. Electromagnetic radiation can interact with matter in many ways. Some types are absorbed easily while others pass through great thicknesses of material. In some circumstances, as in the case of visible light and mirrors, the radiation may be reflected. Absorption can be beneficial – as in the use of light by plants for photosynthesis or the absorption of visible light by pigments in our eyes which enables us to see. Energy absorption may also be harmful however – too much heat absorbed causes burning while other types of electromagnetic radiation can cause cancer.

Electromagnetic radiation can be thought of as travelling in waves, like the waves which spread outwards when a stone is dropped into a pool. The distance between the crests of successive waves is known as the **wavelength** and the number of waves passing a point per second is known as the **frequency**. Each type of radiation has a characteristic frequency or wavelength range – red light, for instance, has wavelengths of around 700 thousand millionths of a metre, while FM radio transmissions are broadcast on frequencies in the 87–108 MHz (million cycles per second) range. If you multiply the wavelength of an electromagnetic wave by its frequency you get the speed at which the wave travels – the speed of light. All electromagnetic waves travel at the same speed unless impeded.

Sound

Sound resembles light in some ways – it can be absorbed or reflected and it travels in waves – but it is not a form of electromagnetic radiation. It is still characterised by wavelengths and frequencies, however, and each note has its specific frequency. Sound is really a series of changes in pressure in the air and it requires a medium to carry it – sound cannot be transmitted through a vacuum. As the saying goes, in space no-one can hear you scream!

The energy carried by sound can be harmful since it can damage the delicate mechanism of hearing as well as causing psychological problems. Sound and **vibration** – such as the rumble of heavy lorries transmitted through the ground – can also set up resonance in buildings and other structures. This is because every system, be it the column of air in an organ pipe or the arch of a bridge, has a natural frequency of vibration. If sufficient energy is applied at the right frequency the system can be induced to vibrate in its natural mode. In the case of a structure this can cause it to fall apart, as the inhabitants of Jericho are reported to have learned to their cost. Marching armies have traditionally broken step when crossing bridges in order to prevent this: contemporary damage to bridges comes more from heavy lorries than from soldiers setting up resonance.

BIOLOGY

DNA and evolution

One of the things which distinguishes the earth from the other known planets is the presence of **life**. Despite the enormous variations in size and function of living things, they all have one thing in common: they contain **DNA**. DNA is the genetic blueprint which, in conjunction with the organism's environment, determines the structure and function of each part of the organism and co-ordinates its development. DNA is inherited, with part of it coming from each parent in organisms which reproduce sexually. Before reproduction – and before cell division – the DNA has to duplicate so that copies of the genetic information can be passed on.

If something goes wrong with the inheritance or duplication of DNA in the cell a mutation can result. This may be of no consequence (most mutations are thought to be neutral), it may confer an advantage on the mutant individual, or it may be deleterious. In the last case the individual may die, may be weaker than the normal form or may otherwise be at a disadvantage. Some types of damage to DNA lead to cancer and factors which can cause such damage include radiation and various chemicals.

In any given population there are variations in form and function which occur as a result of the rearrangement of DNA prior to reproduction. Some of these forms will be better fitted than others to survive in the prevailing environment which may favour them because they are, for instance, taller, stronger, or more resistant to heat. The forms with such advantages will tend to be more successful in surviving and breeding and will come to dominate the population. If environmental conditions change, however, a different form will be more favoured and will come to dominate. This is the basis of **evolution** by natural selection and the underlying process can be traced back to inheritance and DNA. Evolution tends to be a slow process, however, and in some instances change may take place too fast for organisms to adapt. On the other hand, the evolution of resistance to insecticides in some insects happens very quickly and can greatly reduce the useful life of a new product. Evolution is quicker in organisms which have short generation times – in a bacterium which reproduces every twenty minutes it can be observed in a few hours. Human evolution takes very much longer.

Nutrition

The energy which living things need mainly comes from their food, although warmth from the sun is needed by many organisms to maintain a suitable internal temperature. With the exception of a few oddities which live in the ocean depths and feed on inorganic chemicals in warm underwater springs, all food consumed by living things can be traced back to the sun. Green plants trap sunlight and use it, with the appropriate catalysts, to convert carbon dioxide and water into sugars: this process is **photosynthesis** and the green pigment in plants, chlorophyll, acts as a catalyst. Animals eat the plants and plant remains are digested by fungi, bacteria and other micro-organisms. The same fate befalls dead animals eventually, and also the wastes they produce during their lifetimes. In this way energy from the sun is recycled through the system in a chemical form, as compounds of carbon. This is the biological part of the **carbon cycle** (see Fig. 1).

Living things do not just need energy: they need minerals and other basic substances as well. Plants need nitrogen compounds to form proteins which are then consumed by other organisms – although they usually have to break the proteins down to amino-acids by digestion before they can be absorbed. The cycling of nitrogen compounds through living things is an important part of the **nitrogen cycle**. Phosphorus is needed by all organisms to assist in energy transfer reactions in the cells while other inorganic substances are needed for making enzymes, chlorophyll, blood, bones and other essential materials.

Living things can be organised into **food chains** and **food webs** for the purposes of study. A food chain consists of a series of organisms any one of which feeds on the one below it and is fed upon by the one above. This is a simplified version of a food web wherein a range of organisms is placed on each level and links are drawn between them according to which feeds on what. Virtually all food webs and food chains are based on photosynthesis by green plants. At the top level occur the large predators such as lions, eagles, sharks and humans. In between are the herbivores and carnivores which are themselves consumed by carnivores further up – such as shrews which are eaten by owls. Decomposer organisms could be considered to be at the top level since they feed upon the remains of the top carnivores but since they also feed upon everything else's remains and provide food for some creatures at much lower levels they are normally considered separately.

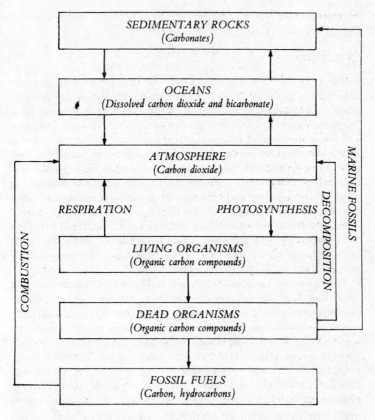

Fig. 1 The main parts of the carbon cycle. (The principal form in which the carbon is present is shown in brackets).

Food webs characterise particular **ecosystems** – an ecosystem consists of the living organisms together with factors in the environment with which they interact such as the soil, air or water in which they live. As a general rule, the more complex an ecosystem is the more stable it is, since diversity enables it to adapt to change. Very simple ecosystems are more prone to disruption. A well established meadow with dozens of different plant species can survive changes in nutrient availability or upsurges in populations of insects or plant diseases much better than an artificial meadow consisting of one strain of rye grass which is heavily fertilised.

Populations

The numbers of organisms in any particular population are determined by several factors, not the least of which is food supply. This, in turn, is determined by factors affecting plant growth such as soil fertility, water and levels of sunlight. A given area of land will have a limit to the amount of plant material it can grow and this will limit the number of herbivores and so on. Many human populations are larger than the carrying capacity of the land on which they live and hence are dependent on imported food.

The interaction between predators and prey is another factor controlling population sizes; a balance between their numbers is established. This is a moving balance – a dynamic equilibrium – which means that numbers of each fluctuate from year to year about average. For instance, populations of owls feeding on mice and voles will increase when mice and voles are plentiful but this will tend to reduce the numbers of mice and voles available. Thus the owl populations will decrease again and the vole and mice populations will recover – and so the process goes on.

The balance between predators and prey is essentially self-regulating but can change in response to external influences. Pesticides can reduce predator numbers, for instance; barn owls have been seriously affected by insecticides as well as by the removal of their nesting sites and this is reported to have led to increases in mice populations in some areas. The use of insecticides in orchards has reduced predator numbers so that the red spider mite, which was previously of little significance, is now a major pest. Natural phenomena such as climatic changes can affect the numbers of organisms living in a particular area and this can also affect the balance.

The biosphere and natural cycles

Living organisms are part of the environment and affect it as well as being affected by it. The term given to the total of living things and the factors directly involved with them is the **biosphere** and this has evolved over millennia since the first signs of life emerged. The climate and atmosphere of the planet have been shaped by life and the changes made have enabled new forms of life to emerge. The first life forms were extremely simple and lived entirely underwater. Gradually these evolved into plants and a major milestone was reached when photosynthetic organisms evolved. This meant that oxygen began to appear in the atmosphere as a by-product of photosynthesis. This had two consequences: it paved the way for **aerobic** respiration which is more efficient at extracting energy from foodstuffs than **anaerobic** (without oxygen) respiration. It also enabled the establishment of the ozone layer which formed when ultra-violet light irradiated the oxygen. This, in turn, provided the shield which enabled life to move out of the water onto land.

At the same time as oxygen was released into the atmosphere, carbon dioxide was being removed. This meant that large quantities of the element carbon became 'fixed' in living things in chemical compounds such as starches, fats, proteins and others. Some of this carbon was released back into the air, as methane or carbon dioxide, when the organisms died and decomposed, but some of it was laid down in swamps and sediments to form coal, oil and natural gas deposits. Removal of carbon dioxide from the atmosphere reduced the temperature of the earth's surface, although this was not the only factor involved. These processes took hundreds of millions of years and fixed carbon was trapped as fossil fuels much more slowly than it is being released by human use of these fuels. As a result of this imbalance, carbon dioxide levels are now rising – see Greenhouse effect in Part 2.

Living organisms remove carbon dioxide from circulation in another way. Marine organisms use carbon dioxide dissolved in sea water to make calcium carbonate, the chemical which makes up their shells. When the organisms die the shells remain (except in some deep ocean environments) and form sediments. These are eventually compressed and formed into rocks such as limestone which can be uplifted by geological processes and exposed at the earth's surface. Large areas of the land are thus the remains of creatures which absorbed carbon dioxide from the sea, and hence from the atmosphere, aeons ago.

The processes described above are part of the carbon cycle, one of many natural cycles by which elements are transferred between living and non-living parts of the environment. These biogeochemical cycles enable levels of nutrients to be renewed and life to continue. If, for instance, nitrogen flowed only one way, through organisms to the external environment without a return phase, soil fertility would have disappeared long ago. Taken over a long enough period and a wide enough area, the carbon, nitrogen and other cycles are broadly self-sustaining if left alone – which is rarely the case. Human interventions have caused serious problems, often by injecting too much of a material in one part of the cycle – like nitrates in rivers or carbon dioxide in the atmosphere.

Substances not essential to life also move in cycles. Lead, for instance, is released in some volcanic emissions and is weathered from rocks exposed at the surface by the action of water and wind. It may spend some time in solution and suspension in river water or in soils, perhaps passing through a food web. Ultimately, the metal will end up in sediments which will be converted into rocks again. Over millions of years, these may be uplifted and exposed at the surface and the cycle will repeat.

It would be wrong to consider the earth as a static system in its untouched state. It has spent the past 4,500 million years or so evolving and will continue to do so. The continents are moving and, over a timescale of hundreds of millions of years, the location of any point on the globe will change drastically as will the climate. Some parts of the land surface will disappear as one piece of the earth's crust moves under the edge of another and is reabsorbed into the planet's interior. New continents will also arise and new rocks will form from sediments and volcanic action. Life, too, is constantly changing and the pattern of species inhabiting the earth in 200 million years time will undoubtedly differ from those living now. These changes are simply a continuation of processes which, as far as we can tell, have operated since the earth cooled down.

Against this background we must consider human impacts. The rates at which human activities are mobilising additional quantities of toxic metals, for instance, frequently approach and sometimes exceed the natural rates. This means that concentrations in parts of the biosphere are much greater than would be present naturally and mechanisms which organisms have evolved to cope with them are becoming overloaded. Similarly, the addition of large quantities of nutrients causes pollution on a localised basis and emissions of certain gases seem to be leading to climatic changes. A further point is that some of the materials discharged by humans have no counterparts in nature. This means that they may not be broken down readily and may have unpredictable effects on natural cycles.

The difference between natural changes in the environment and those brought about by humans is principally one of speed. Fossil fuel deposits took hundreds of millions of years to accumulate. Humanity has evolved over the past three million years or so while an ice age develops and declines over some tens of thousands of years. By contrast, humans have raised the level of carbon dioxide in the atmosphere drastically in less than a couple of centuries and damaged the ozone layer with materials which were barely heard of fifty years ago. Never before has a single species been capable of affecting global systems so profoundly in so short a time.

2: POLLUTION PATHWAYS

Once an effluent is discharged from a pipe or a chimney, or a substance is dumped on land, it can follow many pathways and there have been numerous instances of pollution occurring in unexpected ways, far from the original point of release. This is a particular problem with persistent pollutants since they do not break down readily and can be re-concentrated by living organisms, sometimes to dangerous levels. Nevertheless, the principle of 'dilute and disperse' remains the basis of much pollution control and this does work in many instances providing that the pollutants break down and the natural systems into which they are discharged are not overloaded.

Air pollution

The use of tall chimneys to emit air pollutants is an example of dilute and disperse, the aim being to ensure that by the time the emissions reach the ground they are too dilute to do any harm. This is not as simple as it sounds since a variety of factors such as chimney height, the shape of the land, weather conditions and the nature of the emitted effluent all determine what happens.

Normally, the air in the lower levels of the atmosphere gets cooler as the height above ground increases. This means that warm gases, from a furnace chimney say, will be lighter (less dense) than the surrounding air and will continue to rise to a considerable height. In perfectly still air they will form a narrow cone whose apex is on the top of the chimney. In reality, the air is rarely still so the gases will form a cone-shaped horizontal plume stretching out from the top of the stack with the cone pointing in the direction of the prevailing wind. Pollutants from the chimney will reach the ground at the point where the edge of the cone first touches it. As the gases spread out in the cone they become more dilute so the broader the base of the cone

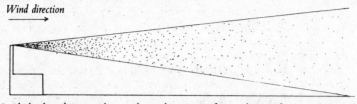

In idealised conditions with a steady wind, emissions from a chimney form a cone shape. When the pollutants reach the ground they have been diluted to a much lower concentration than their original level. In practice, the situation is more complicated, but the diagram shows the general principle.

Fig. 2   How emissions from a chimney spread out.

is, the lower the concentrations of pollutants when they finally hit the ground. This, in turn, depends on the height of the chimney since a tall chimney generates a cone with a broader base than does a short one (see Fig. 2).

Of course, these are ideal conditions and several factors can cause pollutant concentrations to be higher closer to the chimney than expected. The nature of the emitted gases is important – if they are cold and damp they tend to be denser than the surrounding air and simply fall to the ground close to the chimney. This may occur when fumes from an incinerator or power station are scrubbed in a wet process to remove toxic materials. In such instances, the gases should be re-heated before discharge to ensure that they have sufficient 'lift'.

The shape of the land surface is also important. If the land slopes away from the chimney then the cone can spread out more before it touches the ground and thus the pollutants are better diluted. Conversely, if it slopes upwards the plume will hit the ground sooner and concentrations will be higher. Irregularities in the land can cause changes in the direction of wind flow which may also influence the dispersal of pollutants.

One common bar to the dispersal of air pollutants is known as a **temperature inversion**. Here a layer of warm air sits on top of cool air not far from the ground surface; this means that warm gases rising from a chimney encounter a sort of lid through which they cannot penetrate since the warm air is lighter than the effluent. High concentrations of pollutants can build up beneath an inversion layer, sometimes with fatal results. The great London smog of 1952 occurred when inversion conditions trapped the smoke from London's power stations and domestic fires over the city for several days and the Los Angeles photochemical smog forms because an inversion, coupled with the local geography, prevents traffic fumes from dispersing. In the case of Los Angeles a breeze blowing inland helps to keep a lid over the city.

The direction and strength of the prevailing wind can also influence ground-level concentrations of pollutants. If the wind speed is much greater than the speed at which gases are discharged from a chimney the fumes can be sucked down behind the chimney stack into a low pressure zone and their rise can be prevented. Upward air movements – the thermals used by glider pilots to stay aloft – are caused when air near the ground is warmed by the sun and floats above cooler air, forming convection cells. These can improve the mixing of the effluent gases with the surrounding air and may also cause the plume to take on a wavy shape. Strong, steady winds will obviously carry discharges long distances, especially if the gases are released at a great height: emissions of sulphur dioxide from British power stations, discharged from tall chimneys, are carried to Scandinavia by steady winds high above the ground, with dispersion being limited. Thus Britain contributes to Sweden's acid rain problem.

The contents of an effluent plume may not stay the same as it travels, since chemical and physical processes will alter the nature of the emissions. A plume containing water vapour will cool and the water will condense, falling out as droplets. These droplets can dissolve some gases or wash out particles of dust or grit. The action of oxygen, sunlight and water vapour can bring about chemical changes – sulphur dioxide, for instance, is oxidised by oxygen to sulphur trioxide which dissolves readily in water to form sulphuric acid in rain. By the time pollutants reach the ground they may be rather different in nature from those which were originally emitted, particularly if they have travelled a long way.

It is possible to predict the concentrations of pollutants at ground level using computer models of emission rates, wind speeds and chimney heights. Thus the optimum height for a chimney or, alternatively, the maximum amount of material which should be emitted to ensure that air quality standards are met, can be calculated.

So far, emissions from tall chimneys have been considered, but many air pollutants are emitted from sources at ground level such as motor vehicles. In cities with tall buildings and narrow streets dispersion is reduced and high levels of pollution can build up. This is particularly the case when inversion conditions place a lid over the city. It is possible to predict the total average pollution levels in a city as a whole if it is known how many vehicles, heating appliances and industrial premises emit materials, and the quantities of each pollutant which they produce, but it is much more difficult to predict levels in smaller areas of the city where very high levels may accumulate. Where winds blow polluted air from cities it is possible to track the movement of this air for some distance downwind – high concentrations of ozone downwind of London have been detected as a result of traffic fumes emitted in the city and subsequent reactions in the sunlight.

The dispersal of a pollutant and its dilution are not necessarily the end of the story. If the ecosystem into which the diluted pollutants fall is very sensitive, harm may still occur as when acid rain falls into Scandinavian lakes. If the pollutant is persistent then more widespread harm can result, since food webs can become contaminated and bioaccumulation (q.v.) may occur.

The case of lead from motor vehicles is a good example of dilution and dispersion being an unacceptable means of control. Dr Robert Stephens, then Reader in Chemistry at Birmingham University, once commented that it would be difficult to imagine a more effective device for producing an easily dispersed cloud of lead particles than a motor car running on leaded petrol. Furthermore he pointed out that the road system carrying such cars could hardly be bettered as a means of distributing this lead to areas where crops are grown and people live.

Lead emitted by vehicles obviously contaminates the air, but levels in the atmosphere in most places are not acutely dangerous and do not account for the greater part of lead exposure. But as the lead settles out of the air it contaminates roadside dust, falls onto surfaces where food is prepared and eaten and also contaminates crops where they grow. Soil, too, receives lead from the air and some of this can be taken up by food plants, while very small amounts also end up in reservoirs for drinking water. Thus lead emitted into the air can reach us by many routes, notably the food we eat and the dust which children swallow when they put dirty fingers in their mouths, as well as by direct inhalation.

There are many examples of air pollutants causing harm by other routes. Horses near a smelter near Bristol were poisoned by lead falling out onto the grass in the early 1970s, radioactivity from the Chernobyl nuclear accident in 1986 contaminated food chains throughout northern Europe, and dioxins emitted by municipal incinerators have recently been found in milk as a result of cows eating contaminated grass. These examples emphasise the need to consider the environment as a whole and not as a series of unrelated compartments.

Water pollution

When pollutants are discharged into water, similar arguments apply, although the mechanisms of dispersion and dilution may be different. A river differs from the atmosphere in one very important respect in that it contains – or should contain – living things which can destroy polluting substances. However, these living organisms are often harmed by pollutants themselves so it is important to consider a river or other water body as a living entity which must not be overtaxed by pollution.

An effluent discharged into a slow-moving river may take some time to become diluted – a plume similar to an air pollution plume can often be seen downstream from a discharge as the effluent spreads out slowly. In a turbulent river mixing is obviously much quicker and dilution is achieved sooner. Where the effluent is much denser, or much less dense, than the receiving water, mixing may be inhibited as one mass of water may simply sit on top of the other especially if there is little turbulence. In such conditions mixing is by a process of diffusion – the random motion of individual molecules and ions – and this can be very slow. Standards for effluent discharges are often set on the basis of adequate dilution being achieved fairly quickly and turbulence is taken into account.

Large particles of solid material such as grit will normally settle out of the stream fairly quickly unless there is considerable turbulence, in which case they may be carried long distances. Finer particles can be carried a long way in even the smoothest of streams before settling out. Dissolved materials do not, of course, settle out but chemical reactions may occur which cause them to precipitate; some iron compounds, for instance, are oxidised and form orange-red deposits on the stream bottom. Liquids which do not mix with water will move in a separate layer, usually on the surface; oil is the commonest example.

Much of the material discharged to rivers is biological in origin and can be used as food by micro-organisms. Where sufficient oxygen is present in the water decomposition is by aerobic organisms but where the oxygen is used up anaerobic decomposition takes place (see Chapter 3 for the effects of pollution on river life). Whichever process operates, the organic matter is broken down into simpler compounds although the nature and environmental effects of these breakdown products will vary.

The fate of persistent materials depends on their form and the nature of the receiving river. Many metals are poorly soluble in river water but travel as particles or as ions bound onto other particles. These may settle out and remain in river and lake sediments, ultimately to be removed from the biosphere. In other cases these particles may be absorbed by animals which feed by filtering materials out of the water. These filter-feeders can accumulate high levels of heavy metals; indeed mussels are sometimes used as indicators of marine metal pollution. Once absorbed by one species, persistent materials can be passed through food webs and affect other species, even humans. In parts of the Bristol Channel, pollution from industrial activities and run-off from rocks containing heavy metals have led one environmental health department to set maximum limits for the consumption of locally caught shellfish because of their cadmium content.

Modern water treatment plants can remove many of the pollutants present in water before it is supplied for drinking purposes. Some materials still get through, however, including nitrates, pesticide residues, some drug residues and – in a few cases – micro-organisms. The discharge of a material into a river does not mean that it is gone forever, particularly when a river has several towns along its length each of which discharges sewage effluents and abstracts water for drinking. The saying that a glass of water consumed in London has already passed through several kidneys underlines this effectively.

Land pollution

Pollutants discharged onto the land also tend not to disappear permanently. There are many pathways by which they can re-enter the biosphere, the obvious one being direct contact with people and other species when contaminated soil is excavated, built on or simply played in. Some of these problems are described in Part 2 but there are some general principles worth discussing here.

As with water pollution, there is a difference in the fates of degradable and non-degradable pollutants. Materials which can act as food for micro-organisms, such as some of the components of domestic waste, wastes from the food processing industries and some industrial chemicals, may be broken down and converted into fairly innocuous materials – although methane (q.v.) produced in rubbish tips is far from harmless, either locally or globally. This means that some types of contamination will disappear, although this can take a long time and a substantial hazard may exist in the meantime.

Persistent materials, however, present a long-term hazard. Percolating rainfall may carry some of these away from the surface layers of the soil, or wash them into streams, but this means that the problem is spread rather than solved. The contamination of groundwater by materials dumped at the surface is a real risk which has become a reality in many places – for instance in Oxfordshire solvent wastes dumped at Harwell were found in 1990 to be threatening a water supply – and run-off into streams can cause serious pollution. Measures can be taken to reduce the rate at which deposited materials move into water but, over centuries, move they will.

Movement through the ground depends on the permeability of the rocks and soil. Clay has a very low permeability and is very good at retaining deposited materials. It is often used to line landfill sites. Limestone, however, is much more permeable and is often fissured and split. Where there are cracks in the rock, flow is very quick and underground water systems can be contaminated.

This phenomenon accounted for Bristol's water supply turning green in the late 1960s. Sheep farmers in the Mendip Hills, to the south of the city, used to discard sheep dips containing the insecticide dieldrin into natural holes in the ground which led to underground streams and caves. For some time these accumulated but a minor earth movement or similar disturbance suddenly released the residues of these chemicals into new streams which led to a major reservoir. The insecticides were very dilute but still sufficiently powerful to kill off the very small animals which fed on algae in the lake. As a result, the populations of the algae exploded and even after filtration some of the water was still green.

Food chains can obviously be contaminated by materials deposited on land – the example of lead has been mentioned. Some crops are selective about the substances which they absorb through their roots but edible parts above the ground may be contaminated with soil and dust containing toxic materials. Woodlice appear to be efficient at concentrating heavy metals while mosses also retain these substances on their surfaces – bags of moss suspended in trees have been used to monitor air pollution by metals.

Volatile (easily vaporised) materials dumped on land will evaporate unless measures are taken to prevent this – indeed some types of land contamination can be cleaned up by blowing air or steam through the soil to remove volatile substances. Solvents are most likely to evaporate but other materials, such as PCBs, have been known to volatilise from landfills, albeit slowly. The rate at which volatile materials evaporate depends on the prevailing temperature and a property of the material known as its vapour pressure. Materials with a high vapour pressure evaporate very easily.

As a substance moves along a pollution pathway its nature may change. It may be biodegraded, as described above, or it may be converted into another persistent form. In some cases this may be less toxic than the original material but in others the toxicity may be retained or even enhanced. The insecticide aldrin, for instance, is converted in the environment into the equally poisonous dieldrin while DDT is transformed into DDE, the compound which was responsible for eggshell thinning in birds (see Chapter 3).

Pollutants move through the environment in a complex manner by a wide variety of routes. Predicting the ultimate fate of a discharged material is by no means easy, particularly if the substance is persistent. Natural systems constantly produce surprises when they are called upon to handle pollutants. This emphasises the need for very careful consideration before a new material is released into the environment and for vigilance once any new discharge has commenced.

3: THE EFFECTS OF POLLUTION

The effects of pollution are many and varied, ranging from the toxicological to the economic – not that these two categories are necessarily always separate. Materials, plants, wild animals and people can all be damaged by the effects of pollutants, sometimes obviously and sometimes in subtle ways. Specific examples, such as acid rain and lead poisoning, are discussed in Part 2, but in this chapter the general principles are presented.

How pollutants penetrate the body

The pollution issues which make the headlines are usually those which involve harm to humans or other animals, so the effects of pollutants on animals will be considered first. Before a toxic material can do harm it must enter or affect the surface of the body and this it can do in several ways. The skin – or a similar structure in other animals – is there to form a barrier between the body and the outside world and it is generally very good at preventing the ingress of harmful materials. However skin tissues are vulnerable to attack by some substances, such as acids and strong alkalis, and irritation can be caused by a variety of others. Detergents, for instance, remove oils from the skin which becomes dry and sore; degreasing solvents have a similar effect. Skin diseases can be caused by some pollutants – chloracne (q.v.) results from exposure to dioxins (q.v.) and dibenzofurans – while some people are allergic to a variety of substances and their skin can react dramatically to even limited exposure.

The eyes and the mucous membranes are very sensitive tissues at the surface of the body and these are easily irritated by pollutants. The mucous membranes line the mouth and nose, among other structures, and do not present as good a barrier to toxic materials as does the normal skin; nicotine in snuff and tobacco smoke, for instance, passes into the blood through the mucous membranes. There have been cases of people taking a poison – e.g. the herbicide paraquat – into the mouth and spitting it out without swallowing, yet still absorbing a lethal dose through the lining of the mouth. Ozone is a powerful irritant to the eyes and mucous membranes and some solvent vapours have a similar, though less pronounced, effect.

The skin is not a perfect barrier and some materials can pass through it, notably those which are soluble in oils. Percutaneous (through-the-skin) poisoning can occur with some pesticides, organic metal compounds such as tetra-ethyl lead and methyl mercury, and some solvents. Even inorganic lead compounds can move through the skin to some extent although this is not a major pathway. If the skin is damaged, by abrasion or chemicals, it is easier for toxic materials to pass through it. Once through the skin the substance can be carried in the bloodstream to a sensitive organ, such as the brain in the case of organic metal compounds, and/or stored in fatty tissues.

Inhalation

A second route of exposure to toxic materials is inhalation (in water-living organisms, taking in a material dissolved in water through the gills is roughly analogous). The lung has an enormous surface area which means that gases and fine particles can be absorbed very efficiently once they reach the depths of the organ. The abundant blood supply in the lung means that transport of absorbed materials to other parts of the body is very swift. Large particles are filtered out to a greater or lesser extent by hairs in the nose or are trapped in the tubes leading into the lungs but very fine particles and gases penetrate deeply. Some particles are removed from the lungs in the stream of mucus which is set up in the upper tubes but this is not completely effective and many particles remain, hence the blackened appearance of a smoker's lungs and the retention of potentially lethal fibres by people exposed to asbestos. Coughing is a mechanism by which the lungs clear themselves of foreign material but this, too, is not completely effective.

Inhalation is the main route by which air pollutants are absorbed but it is not the only one – see Chapter 2. It is a very efficient means of transferring a toxic material into the bloodstream and hence to sensitive organs. Lung tissues are also sensitive themselves and a variety of materials can irritate them. Furthermore, carcinogenic particles lodging in the lung can induce cancers if not ejected.

Ingestion

The other obvious route by which toxic materials are taken into the body is by ingestion – swallowing. The purpose of the gut is to enable the body to absorb foreign materials and in most instances it does not discriminate between toxic substances and foods. Some substances will irritate the stomach and produce vomiting so that the poison is ejected – concentrated solutions of some cadmium compounds will have this effect, for instance – but many are absorbed readily as long as they are soluble in the fluids which are secreted by the various parts of the digestive tract. Some substances may pass through without being absorbed to any great degree; mercury metal, for instance, will not dissolve in the digestive juices, although some mercury compounds can pass easily through the gut wall into the bloodstream with deadly results. The diarrhoea produced by some poisons is an attempt by the body to get rid of harmful material.

Pollutants in the body

Once absorbed by the body, a variety of fates may befall a pollutant. The body will get rid of some materials by excreting them in the urine or breath, sometimes unchanged or sometimes in an altered form. Sweat and even milk are other routes by which toxic materials leave the body. Some human milk in the USA in the 1960s contained so much DDT that it would have been illegal to sell it if it had been cows' milk. The liver is responsible for detoxifying many unwanted compounds and some materials may be broken down here before they can do any serious harm. Metals sometimes accumulate in the liver, bound in a form which keeps them out of circulation, although they cannot be broken down. In some instances, metabolism (the name given to the vast array of biochemical processes operating in an organism) may convert the absorbed material to a more harmful one. Some insecticides are metabolised into the active substance by the target; hence toxicity can be increased once a material is absorbed.

Substances which dissolve in fat tend to build up in fatty tissues and a balance is maintained between the concentration of material in the bloodstream and the concentration dissolved in fat. If the level in the blood rises, as when more is taken into the body, more goes into the fat; if it drops, some of the stored substance leaves the fat. If the fat in store is suddenly depleted it can release a pulse of toxic material into the bloodstream; someone who has been exposed to DDT for a long period, for instance, will suffer increased levels of the insecticide in the blood if they go on a crash diet. Fat soluble materials can be particularly worrying if they affect the nerves and brain since nerves are surrounded by a fatty sheath in which these materials may accumulate. This is why tetraethyl lead, an organic and fat-soluble lead compound, is much more toxic than inorganic lead.

Bones, too, can store foreign materials. Anything which behaves in the body like calcium, a major ingredient of bone, can be accumulated in bones and such substances include lead, cadmium and strontium. The radioactive isotope strontium-90 was released in large quantities by nuclear weapons tests in the 1950s and caused widespread contamination around the globe. As it behaved in a similar manner to calcium it appeared in milk (which is rich in calcium) and was stored in bones. Teeth also store these materials and a record of a child's exposure to lead can be obtained by analysing shed milk teeth. This is a much more reliable indicator of total lead exposure than blood lead levels which fluctuate according to the amount of lead currently being absorbed.

Substances which build up in the body until they cause harm are known as cumulative poisons. In most cases the materials concerned are not broken down in the body and are stored in tissues which they can harm; heavy metals are examples. Sometimes a chemical may accumulate if its breakdown rate is slow and cannot keep up with the rate at which the substance is absorbed. Methanol, a toxic solvent, is a case in point since daily doses of this substance will lead to a build-up in the body and damage can then occur. A single small dose, or a series of smaller doses, can be metabolised without serious harm.

Acute poisoning

Where a poison produces immediate symptoms its effects are said to be acute. The intensity of these effects will vary according to the dose, the individual (since all individuals respond differently), and the presence or absence of other materials in the body. Depletion of the body's stores of some materials can make an individual more vulnerable to some poisons. Acute effects from pollution in humans are generally fairly obvious; wheezing and eye irritation from high levels of ozone in smog, vomiting from contaminated water in the Camelford incident (see Aluminium) and collapse and death following the Bhopal (q.v.) chemicals leak are clear examples. Acute poisoning by pollutants occurs in other animals as well, notably fish in poisoned rivers and a variety of species caught in pesticide sprays.

The effects of acute poisoning may pass if they are not fatal but in some instances lasting damage may be caused. A moderate exposure to carbon monoxide causes drowsiness and headache which will clear once the victim is placed in clean air: a larger dose can cause irreparable brain damage. Drinking alcohol can have similar effects.

Chronic poisoning

Acute poisonings make the headlines but less dramatic though probably more common are chronic poisonings. Chronic poisoning takes place over a long period and usually results from repeated doses of a poison or continuous exposure to it. The poison itself may be cumulative or it may be that the effects of each dose build up as every small dose does a bit more damage, until a threshold of serious harm is passed. Low-level exposure to mercury is a classic example. Frequent exposure to mercury compounds gradually caused brain damage and mental illness in hat makers in the last century, whence the expression 'mad as a hatter'.

Chronic poisoning can be difficult to detect at first since the symptoms may be confused with everyday malaises, viral infections or an unhealthy lifestyle. Again, mercury provides a good example since low-level chronic mercury poisoning produces symptoms of irritability, tiredness, confusion, poor muscular control (the victim's handwriting becomes difficult to read) and headache. All of these could be ascribed to other causes and mercury exposure would not be suspected by a doctor unless it was known that the sufferers, who include dental technicians and nurses, worked with mercury.

It is also difficult to determine the chronic effects of pollution on populations. Some studies have been carried out in an attempt to link pollution levels with infant mortality and other indicators of general community health. Unfortunately, it is almost impossible to find two comparable populations which differ only in the amount of pollution to which they are exposed. There are usually other differences as well since areas suffering the highest levels of pollution tend to have higher levels of unemployment, poverty and other social problems too. Nevertheless, a tentative link was found in a study carried out in Bristol and further research is under way. It is certainly known that levels of lung cancer are higher in cities and this may be linked in part to pollution.

In some grossly polluted areas the effects of pollutants on the health of local people are glaringly obvious. Newspaper reports from the Romanian town of Copsa Mica reveal illness on a wide scale which is obviously caused, in a large part, by the horrendous levels of air pollution from the unregulated factories operating in the area. Some of the illnesses reported are acute but others – such as lead poisoning – are chronic. Such obvious problems are rare in the West but chronic effects of pollution on health, particularly on the health of vulnerable groups such as the elderly, people with lung and heart diseases and the very young, undoubtedly occur.

Reproductive effects

One aspect of chronic toxicity which is becoming increasingly important in the pollution sphere is the effect of chemicals on reproduction. It has long been known that some pesticides can affect the breeding success of birds by causing their eggshells to be thinner. Now it seems that mammals may be affected. Dieldrin has been found to impair the fertility of otters and PCBs are thought to have caused reductions in the populations of Dutch seals at the mouth of the Rhine by damaging their fertility. PCBs seem to be affecting British otters although dieldrin is now less of a problem. Pollution has not been shown to impair human fertility on a wide scale so far, although some farmworkers using large quantities of pesticides in the 1960s were thought to suffer from impotence as a result of chemical exposure. There is evidence to suggest that the pesticide pentachlorophenol can reduce the sperm count in humans and more work needs to be done in this field.

It has long been known that exposure to certain chemicals in pregnancy can cause birth defects in the child. Substances which cause this effect are known as teratogens. Some organic mercury compounds are teratogenic in humans while a range of other materials, including some pesticides and dioxins, behaves this way in animals.

Cancers

Materials which are teratogenic may also be carcinogenic – capable of causing cancer. Cancer is the out-of-control proliferation of cells in the body and this condition can be triggered by a variety of stimuli including radiation, viruses and toxic substances. Cancers can take a long time to appear – up to 40 years in some cases – and many forms cannot be cured especially if they are detected late. Environmental carcinogens include asbestos, plutonium, polycyclic aromatic hydro-carbons and benzene as well as several types of pesticide.

Identifying a carcinogen is not easy. Laboratory studies can be carried out in which a suspect substance is tested on animals and the incidence (number of cases) of cancer is noted. This is compared with the incidence of cancer in a control group of animals who are matched very carefully with the experimental group and treated in exactly the same way apart from the exposure to the chemical. If the results show an increase in the number of cases of cancer when the chemical is applied it can be inferred that the material is carcinogenic in that species. It does not prove, although it suggests, that it may be carcinogenic in another species.

Similarly, the absence of an increase in cancer cases does not mean that the substance is not carcinogenic in other animals. Testing chemicals on animals to determine their safety in humans has an inherent weakness.

Population studies can be carried out where a group of people exposed to a material by virtue of their work, lifestyle or location are compared with a similar group of people elsewhere and the two rates of cancer are established. This is difficult to do since many factors can complicate the exercise including exposure to materials other than the one under study. Sometimes suitable groups can be found – the link between smoking and cancer was first established by analysing smoking patterns among doctors and monitoring causes of death – but it is rare to find groups similar in every respect apart from exposure to the chemical.

If a chemical is found to be carcinogenic (or teratogenic) it does not mean that every individual exposed to it will necessarily suffer the condition. Apart from variations in dose, each individual will react differently to the same dose – some people may detoxify the substance more effectively, others may be more resistant to cancer anyway, and so on. The potency of a carcinogen is often expressed in terms of the number of excess cancer cases which exposure at a given dose rate will cause in a population. This can only give numbers; it cannot identify those individuals who are most at risk and nor can it identify a safe dose of a carcinogen. Alternatively, a given dose can be said to increase an individual's chances of developing cancer by a certain amount and this can be compared with the risk from other exposures to carcinogens such as medical X-rays or tobacco smoke.

Thresholds of safety

There is some dispute over the interpretation of this type of data. Some people believe that there is a threshold of exposure, below which there will be no increase in the incidence of cancer. Others believe that any exposure to a carcinogenic material increases the individual's risk of cancer and, if the exposed population is large enough, will cause extra cases of the disease.

Calculating a safe level of exposure to any material is a risky business since groups of people and individuals within those groups vary so much. Children, for instance, are at greater risk from air pollutants than adults since they breathe proportionately more air in relation to their body weight. This means that they can absorb a relatively larger dose from a given concentration in the air. There is also evidence to suggest that children retain more of a swallowed dose of some toxic materials than do adults as they absorb them more easily. Elderly people, and those with heart or lung diseases, are also particularly vulnerable to air pollution since anything putting extra stress on their breathing, such as irritant fumes which cause coughing, can be dangerous. Pregnant women – or, rather, their unborn children – also make up a high-risk group and evidence is emerging to show that couples wishing to conceive should also take special care in avoiding exposure to toxic chemicals.

Levels of pollutants to which people in industry may be exposed are decided by the Health and Safety Executive and published annually in guidance notes. These are calculated on the basis of exposure over a normal working day, although there may be short-term exposure limits for some particularly hazardous substances. It is normally assumed that the exposed person will be a fit male between the ages of 17 and 60, although special considerations apply to women in the case of lead. The exposure limits are not designed to protect the general population which contains people younger and older than the industrial workforce and also those who may be ill. Exposure to an air pollutant in the community may also be for much longer than 8 hours per day. A rough idea of an acceptable limit for the community is sometimes obtained by dividing the industrial limit by 30 or 40 although some people feel that this is not strict enough.

Sensitivity in individuals

Some individuals are much more sensitive than normal to particular substances because they suffer allergic reactions. An allergy is an over-reaction of the body's defences against a particular substance and usually occurs after an initial exposure to the material which causes sensitisation. An allergy can take several forms depending on the agent involved and its exposure route. Inhalation of isocyanate vapours, for instance, can cause several asthma-like symptoms, while nickel on the skin can bring allergic people out in a rash. Allergy is not confined to pollutants; hay fever is an allergic response to pollen while various foods, from strawberries to shellfish, cause allergies in some people.

Another type of sensitivity concerns people who react adversely to low doses of organophosphorus, and other similar, insecticides. These substances are known as acetylcholinesterase inhibitors because they block the action of an enzyme called acetylcholinesterase. This is important in the transmission of nerve impulses and if it is inhibited to any great degree the nervous system fails – this is the principle upon which military nerve gases work. Some people naturally have low activities of this enzyme which means that comparatively small doses of the insecticides will cause low-level symptoms such as headache, tremor, twitching and weakness. Such low levels would not normally harm the average individual but sensitive people are advised to avoid contact with these compounds.

Many poisons work by blocking the activities of enzymes. Enzymes are biological catalysts which control all the myriad biochemical reactions which occur within a living organism. If these enzymes are inhibited then the processes which they control may carry on out of control. Some herbicides have this effect and someone poisoned with them becomes very hot as food is burned up at too fast a rate without its energy being utilised. In other cases the reactions may be slowed down or may not take place at all. Sometimes the presence of low-level poisoning can be detected by monitoring the products of particular chemical reactions in the body rather than measuring the pollutant itself. There is a crude but simple test for lead exposure which relies on this principle and involves monitoring a substance in urine.

The effect of synergism

The effects of particular pollutants are usually considered in isolation but that is not always the way in which they occur. Frequently, an organism may be exposed to several different substances at once and their effects may combine. Sometimes this combination may be beneficial to the organisms – elevated levels of zinc, for instance, can protect against the effects of lead – but adverse combinations can occur. Where the effect of a mixture is greater than the sum of the effects of the individual components, the result is synergism. For instance, smoke irritates the lungs as does sulphur dioxide, but a combination of the two irritates much more than the sum of the effects from each pollutant. Nitric oxide may have a synergistic effect with carbon monoxide on the ability of the blood to carry oxygen. The interaction of pollutants in this manner has not been studied in great detail and much work needs to be done in this area.

Plants and pollution

The discussion so far has focused mainly on animals but plants are susceptible to pollutants as well. Plants do not have such sophisticated barriers against unwanted substances as do animals, although their roots can be slightly selective about what they absorb. Plants are simpler, too, and damage to them is often the result of gross harm to leaves and tissues rather than interference with delicate enzyme systems.

Plants, with the exception of fungi, depend on light for their energy supply; hence anything which cuts the amount of light reaching their leaves can reduce their growth. Thus plants grown in areas subject to heavy smoke pollution are likely to be weaker and grow less than comparable plants in cleaner areas. Two factors operate here: dirty air lets less sunlight through to the leaves and the leaves themselves may be covered by a layer of dirt which further blocks the sun.

Heavy smoke pollution can also block the tiny pores – called stomata – which let air into the plants and enable water vapour to leave it. This can effectively suffocate the plant. Of course, toxic gases can also enter via the stomata and damage the plant's tissues; sulphur dioxide is one such gas, as is ozone which is extremely harmful to plants.

Some plants have particular sensitivities to certain pollutants – gladioli, for instance, are very susceptible to damage by fluorides. This can be exploited as a means of monitoring pollution without expensive apparatus. By recording the distribution of lichens, which are colonies of algae and fungi growing together, estimates can be made of the levels of sulphur dioxide in the air since different species of lichen can tolerate different levels of the gas.

Particular patterns of scorching, pigmentation and spotting on leaves are characteristic of certain pollutants. Sulphur dioxide and chlorine, for instance, both cause bleaching while mercury causes yellowing and brown spotting. Some pollutants may mimic the effects of natural factors, such as water shortage, and it may be unclear whether or not pollution is involved in a problem. In many instances, plants may be exposed to a whole range of pollutants so synergistic and other combinations of effects may occur.

The use of herbicides has sometimes caused air pollution harmful to plants. Some of these materials are quite volatile and on hot days with gentle winds significant quantities have evaporated after spraying and have drifted into neighbouring fields. Serious crop damage has resulted and was a major problem in the Vale of Evesham in the early 1980s. Many herbicides work like plant hormones in that small quantities absorbed by the leaves are transported to growing regions where they cause over-stimulation. The plant then grows rapidly and in a highly distorted manner, eventually dying.

Plants can be affected by pollutants in the soil as well as those in the air. Toxic materials such as creosote and other wood preservatives can prevent or stunt plant growth, as can excess levels of heavy metals such as zinc and copper. This is one of the reasons why the use of sewage sludge, which contains these metals, as fertiliser is sometimes limited. Acid rain is thought to leach important plant nutrients from the soil and thus contribute to forest damage and also to mobilise toxic levels of aluminium in some cases. Acidification of soils can also inhibit the activity of bacteria which are important in cycling nutrients within the soil. Plants grown on top of old rubbish tips are sometimes deprived of another vital resource – air. Roots need air and this is normally obtained from the spaces between soil particles. Methane generated by the decomposition of organic matter in rubbish sometimes replaces this air and plants then suffocate. Trees grown over such tips thus grow only to a certain height and then die as their roots grow normally until they reach a zone rich in methane or some other toxic material, at which point they expire.

Damage to ecosystems

Pollution can obviously affect individual organisms and species but a heavy pollution load can have more general effects on an ecosystem as a whole. A common effect is to reduce the diversity in the system as the more susceptible species are killed off or prevented from breeding and the more robust members increase their numbers to fill the gap. Numbers may be reduced overall as well, especially if the pollutants interfere with food production by, for example, inhibiting photo-synthesis. Freshwater acidification has reduced both diversity and population numbers especially in Scandinavian lakes. In Britain populations of the dipper, a water bird which feeds on aquatic insects, have been drastically reduced in acidified streams as the bird's food supply has diminished.

The diversity and identities of the species present in a river can provide a good indication of how clean it is. A clean river will contain a wide range of invertebrate species while a grossly polluted waterway will contain very few species. Furthermore, clean rivers will support sensitive species such as dragonfly larvae while very dirty ones will only contain tolerant organisms such as *Tubifex* worms or the charmingly-named rat-tailed maggot, which breathes air through a tube extended to the water surface. These organisms are known as biological indicators and a scale of cleanliness for streams can be established depending on what lives there. They really indicate the state of oxygenation of the water and do not measure chemical pollutants such as pesticide residues or industrial effluents unless they deoxygenate the water.

Deoxygenation is a crucial effect of pollutants discharged into waterways since the vast majority of aquatic life depends on dissolved oxygen for respiration. Any organic material which can be used as food by micro-organisms can cause deoxygenation as they respire and the higher the BOD (q.v.) – Biochemical Oxygen Demand – of the effluent, the worse is the damage it can cause. Some inorganic materials, such as dissolved iron compounds, can also cause deoxygenation and the problem, however caused, is aggravated in warm conditions. This is because less oxygen can dissolve in warm water than in cold. In addition, bacteria reproduce and feed faster in warmer conditions. Hot water discharged from power stations and other cooling systems can thus aggravate pollution problems, a phenomenon known as thermal pollution.

If a deoxygenating effluent is discharged into an otherwise clean river the worst damage will be a little way downstream from the outfall. Oxygen levels will drop sharply and the community of aquatic organisms will change. Further downstream the river will begin to recover as the waste is used up by micro-organisms and the rate at which fresh oxygen enters the water from the air and by photosynthesis in water plants begins to exceed the rate at which it is consumed. The recovery of the river is accelerated if it is turbulent – e.g. running over rapids – since this increases the rate at which oxygen dissolves in the water. Eventually the river will reach its fully oxygenated state again, unless another effluent is discharged into it in the meantime.

Although organic wastes such as sewage are the major pollutants of rivers in many cases, other materials also cause damage to communities. Excessive levels of nutrients cause eutrophication (q.v.) and algal blooms (q.v.) while high levels of suspended solids can choke water animals and inhibit photosynthesis by blocking the passage of light to water plants. Intensely coloured materials such as dye wastes can have the same effect on photosynthesis. If plants cannot make food by using the sun's energy, a large part of the food web will be affected, leaving little other than decomposers. These will eventually disappear as their food supply vanishes. Layers of oil on the surface of the water cause suffocation of life beneath since they prevent the passage of oxygen and a similar problem can occur if the river is covered with a layer of foam.

Pesticides entering a river can cause serious problems since fish are very sensitive to many of these compounds. Insecticides will also kill some of the invertebrates upon which other species feed. Pesticides enter through run-off from fields, by accidental spraying or when empty containers are illegally dumped in streams. Many other chemicals can have toxic effects in waterways, especially to fish which are efficient at extracting materials from the water via their gills.

Land-based ecosystems are in some ways less vulnerable to pollution than waterways since oxygen is not usually in short supply. Nevertheless, whole-community effects can be found when levels of pollution are high. Anything which affects photosynthesis, for instance, affects the base of the food web and can influence the abundance of many different species. The balances between predator and prey populations and between competing species can be disrupted by pollution – the case of barn owls and rodents was mentioned in Chapter 1. One of the consequences of widespread fertiliser use has been the loss of some wild flowers which do not respond to high levels of nitrate and are out-competed by other plants which can use the fertiliser. Herbicide use is another factor in the decline of these species.

Over time, some adaptation to pollution may be possible. The Severn Estuary has been badly polluted by heavy metals for many decades yet it supports a wide range of species in some abundance. The shellfish found there contain high levels of metals but seem to have adapted to their presence. In some instances individual organisms adapt while in others a population may evolve with some protective mechanism. Resistance to pesticides can develop in this way: within any pest population there will be a wide range of susceptibilities to a given compound. When the compound is sprayed it will kill the vulnerable individuals but the more tolerant ones will survive. While they breed the new population will contain a higher proportion of tolerant individuals and, with repeated sprayings of the same chemical, virtually the whole population will become resistant. This has serious economic implications since larger amounts of the chemical, or more expensive materials, are then used to control the pest and in some instances large crop losses occur. Increased chemical use also increases ecological damage. Not all pests develop resistance to pesticides but many insects, fungi and even weeds do. By 1984 some 447 species of insects and mites had developed resistance to one or more insecticides in some populations.

When investigating the effects of pollutants on populations it is important to remember that many factors may be operating. Apart from synergistic effects which are discussed above, a population may be reduced by two or more completely independent processes. The decline of the peregrine falcon in Britain in the 1950s is attributed to the use of pesticides, notably dieldrin. The dieldrin killed peregrines by secondary poisoning; the birds fed on pigeons which had absorbed large amounts of dieldrin through eating treated seeds. This might not have been so serious, however, had the peregrines not also been exposed to DDT. This was metabolised by the birds into DDE, a chemical which causes eggshell thinning in some species resulting in broken eggs in the nest. Breeding success in the peregrine was already affected by DDT when dieldrin was deployed widely as a seed dressing so it may have been the combined effects of the two compounds, acting in completely different ways, which decimated the peregrines.

A final point to remember when studying the effects of chemicals on living populations is that there are wide natural variations in numbers anyway. Determining the size of the 'normal' or unaffected population may be difficult especially for species which breed quickly or are difficult to observe. The decline in the peregrine was comparatively easy to observe since there were good data available from ornithologists dating back to before the insecticides were used. This is rather unusual and trends may be difficult to detect at first in other species.

Global and economic damage

The effects of pollution are not confined to the living world: inanimate objects are harmed as well. Acidic substances in the atmosphere, mainly from fossil fuel combustion, corrode many types of building material such as limestone, concrete and some sandstones. Metals corrode more rapidly in acidic water as well. Buildings are also soiled by smoke and the costs of cleaning them can be enormous. The soiling of other materials such as furnishings and clothing by airborne particles also has an economic cost as they have to be cleaned more frequently. Ozone in the lower atmosphere causes dyes to fade as it is a powerful bleach; it also attacks rubber and some textiles, causing them to weaken and tear more easily.

Economic effects are not confined to damage to materials. Air pollution causes extensive damage to crops as well as to forests. The tourist industry in areas of Scandinavia affected by acid rain has been hit since there are no fish in many of the lakes and similar concerns are being expressed about tourism in the Black Forest. Ironically, the sulphur dioxide which is a major cause of acid rain is actually beneficial to plants in some areas where the soil is naturally low in sulphur. It also helps to keep down some plant diseases.

Transcending all the aspects of pollution described so far is the question of global effects. There are two major areas of concern: damage to the ozone layer and the greenhouse effect, both of which are described in some detail in Part 2. Suffice it to say at this point that gases and vapours released at the surface of the earth by human activities appear to be damaging the layer of ozone in the upper atmosphere which screens out certain harmful radiations from the sun. Serious damage to this layer will lead to large increases in the rate of skin cancer in humans, mutations in other organisms and damage to a wide variety of species including some at the base of food webs. The other global problem, the greenhouse effect, is predicted in some quarters to be even more serious. Gases such as carbon dioxide and methane released by a wide variety of domestic, industrial and agricultural practices trap heat in the earth's atmosphere, a process which is thought to be leading to a generalised global warming. This could have major effects on climate, sea level, agricultural productivity and land use, and just about every ecosystem on the planet.

From this general account it is obvious that the effects of pollution can be either extremely localised or global in extent. They may vary in intensity from minor illness or a slight depletion of oxygen in a river to the four thousand deaths attributed to the 1952 London smog or the destruction of a forest. The solutions to pollution must therefore also be local and global. Every small discharge of heavy metal into a drain adds to the total burden in the environment and every release of carbon dioxide potentially adds to global warming. Individuals, communities and companies can thus do their share in preventing pollution by examining their own activities and modifying them where necessary. But these efforts are not enough on their own: co-ordination and co-operation between countries is essential if the global problems are to be tackled effectively. This takes us out of the scientific sphere and into the realms of politics, although the science must not be forgotten. On past performance the prospects for appropriate and timely action are not reassuring, although there are some glimmers of hope.

Unfortunately, it may be too late simply to wait and see.

PART 2

DIRECTORY
OF POLLUTION
AND POLLUTANTS

ACID RAIN Acid rain – more properly known as acid deposition – is the phenomenon of acidic gases, solutions and particles in the atmosphere causing damage to the environment. The chemicals in question include nitric and sulphuric acids, sulphate particles and, in some instances, dry sulphur dioxide gas. Trees, fish, insects and birds have all suffered from the effects of acid rain and much damage has been done to buildings made of limestone and other acid-soluble rocks.

The problem often starts with burning. Fuels containing sulphur emit sulphur dioxide when they burn. Most combustion processes, especially those operating at high temperatures (as in power stations or vehicle engines), generate nitrogen oxides as well and, like sulphur dioxide, these gases dissolve in water to form acids. Reactions with oxygen in the atmosphere increase the acidity of the gases and their solutions. These emissions can be carried long distances by the wind – tall power station chimneys are very good at dispersing waste gases away from the point of generation, but they still end up somewhere – and their acidity may be deposited hundreds of miles from its origin. Sulphur dioxide has generally been regarded as the main cause of acid rain but this is an oversimplification. In much of Europe sulphur emissions have declined and nitrogen oxides, which are increasing, are more important as a cause of acidification.

Fuel burning is not the only process releasing acidic gases. Metal smelting often produces sulphur dioxide. Nitrogen oxides are emitted from fertiliser plants. The effects of such emissions are likely to be fairly localised as the chimneys through which they are discharged tend to be smaller than power station stacks. Ironically, emissions of ammonia (which is an alkaline gas) can cause acid deposition since it is oxidised in the atmosphere into acidic nitrogen compounds. Ammonia is released from intensive livestock units.

The effects of acid rain are still controversial. It is known that acid deposition is damaging trees although some of the worst damage appears to be the result of acid rain combining with ozone and other factors in the environment such as drought to restrict the growth of and kill trees. Parts of Scandinavia, Germany and Eastern Europe seem to be most badly affected although acid rain damage has been reported in Britain as well. The official government position, and that of the Forestry Commission, is that acid rain is not killing trees in Britain although the Forestry Commission has reported that the health of many British trees is only 'moderate'. Nevertheless, the Third Report of the UK Review Group on Acid Rain concluded that in many parts of the country sensitive ecosystems were at risk from acid rain, the problem being worst in western hilly areas which have higher concentrations of acid in the water. A report produced by the Austrian forecasting group, IIASA, in 1990 concluded that acid rain will cost Europe some £16 billion per year in lost wood production for the next century unless drastic action is taken to curb emissions of sulphur, nitrogen oxides and ammonia.

Exactly how acid deposition damages trees is also unclear. Some effects may be the result of damage to the leaves while others may result from interference with soil nutrients as important mobile elements such as magnesium are washed out of poor quality soils or toxic aluminium is mobilised. Other pollutants, such as ozone, are also involved in some areas and factors such as drought and disease are also relevant. One theory is that the action of acid deposition renders the tree less resistant to the effects of drought and disease. In Scandinavia and Germany the authorities are more convinced than their British counterparts that pollution, and in particular acid rain, is damaging trees.

Lakes and rivers have certainly suffered from acid deposition, a problem which was first documented in Scandinavia. Many lakes in Norway and Sweden have become virtually lifeless as their acidity has increased. This has two effects: (1) invertebrates – e.g. insects – cannot survive in water that is too acid, and (2) as the lakes are located on rocks containing much aluminium (q.v.), this material dissolves in the acid water and poisons fish. The survival of fish eggs is also greatly reduced in acid conditions. Lakes situated on rocks containing alkaline materials – e.g. limestone – are not affected since the alkali neutralises the acid.

Britain has suffered from acidification in lakes and streams in Wales and Scotland and parts of England may be affected as well. Fish populations have been reduced drastically in some instances. Some Scottish lochs and Welsh lakes are now fishless. One consequence has been sharp declines in populations of the dipper, a small stream-living bird which feeds underwater on insects and their larvae.

A particular problem affecting lakes and streams is the acid 'pulse' in springtime. Over the winter, high levels of acidity are stored in deposited snow and when this melts in the spring a sudden surge of acid hits the streams, acutely damaging the ecosystem. Acid rain is not the only cause of acidification in lakes but it is a major one. Water draining from coniferous forests tends to be acidic and commercial afforestation in the UK and abroad can cause problems by mobilising stored compounds in the soil. However, increasing emissions of sulphur dioxide have been correlated with the disappearance of small acid-sensitive organisms from lakes over the past few decades, findings which support the view that pollution is a major factor in the acidification of freshwater. Furthermore, when acidification started particles of soot derived from coal burning were first deposited in sediments.

Where acid in the rain has declined, some lakes are showing evidence of recovery. In some Scottish lakes a reduction in rainfall acidity of 50% has been followed by a reduction in the acidity of their waters of 30%. Further reductions are needed before the lochs can support fish again and in Wales no such reduction in either lake or rainfall acidity has been found. Norwegian lakes are not recovering either, because increases in the amount of nitric acid in the rain have made up for the decline in sulphuric acid.

Groundwater now appears to be under threat as well as surface waters. A report from the British Geological Survey in 1990 warned that acid deposition could mobilise toxic metals from the ground into aquifers. Water supplies in alkaline rocks, such as limestone, will not be affected since the rocks will neutralise the acid; but in sandstones and other lime-free rocks metals such as aluminium may be dissolved and render water supplies unfit to drink without substantial treatment.

Many historic buildings are suffering from the effects of increased acidity in the rain. A report produced for the Department of the

Environment in 1990 stated that cutting sulphur dioxide emissions by 30% would save between £80 million and £160 million on repairs to old buildings and between £9.5 billion and £17 billion in maintenance to new buildings over the next 30 years. Such a cut in emissions is expected to cost about £11 billion.

The main source of sulphur dioxide is power stations. In Britain they produce 70% of the sulphur dioxide emitted and account for half the concentration of this gas in urban areas. Power stations also produce large amounts of nitrogen oxides although motor vehicles are the major source of these in cities. Strategies for controlling acid rain are therefore most often aimed at power stations, although in Germany a major focus in preventing forest damage has been on vehicles with a view to controlling ozone. Catalytic converters (q.v.) have long been fitted to German cars but are only just being introduced in Britain.

Emissions from coal-fired power stations can be controlled in two main ways. The burners can be adjusted to minimise the generation of nitrogen oxides, a move which has been made on an experimental basis in some stations. For sulphur dioxide removal a post-combustion technique is normally required, although it is sometimes possible to reduce the sulphur content of coal by washing it and oil can be refined so that its sulphur content is minimal. Low-sulphur coal can be used instead of the normal fuel, but this normally means that someone else will burn the high sulphur coal so the problem is simply transferred.

Controlling sulphur emissions after combustion usually involves cleaning the gases with a limestone slurry which absorbs the acidic gases. This involves the use of large amounts of limestone, which may be quarried from valuable scenic areas, and generates large amounts of waste products. There are alternative techniques. Fuels can be burned more cleanly in fluidised bed furnaces, in which the burning material is suspended in a bed of sand or similar material supported on a stream of air. Powdered limestone is added to the bed and this traps the sulphur dioxide as it is formed. Other techniques involve less limestone and can result in saleable by-products.

A programme of flue gas desulphurisation was planned by the UK government in the 1980s in response to international and domestic pressure. Much of this was later dropped as it would have affected electricity privatisation plans. Only one or two stations now seem likely to be fitted with this equipment. The new power companies may install additional desulphurisation apparatus but are expected to rely heavily on low-sulphur coal for sulphur dioxide reduction. By the end of 1989 there were over 500 flue gas desulphurisation plants in operation around the world with over 150 more planned.

One control strategy for acid rain reduction which has been overlooked in some quarters is energy conservation. Simply burning less fuel would reduce the emissions of nitrogen and sulphur oxides and makes much more sense than spending money on repairing corroded buildings or, as the Swedes now have to do, adding large quantities of powdered limestone to poisoned lakes to try to repair the damage. Although international action has succeeded in reducing levels of sulphur dioxide in Europe by 20% since 1980, there is still some way to go in dealing with acid rain and further initiatives to cut sulphur dioxide emissions, freeze nitrogen oxide releases and limit volatile hydrocarbons (which generate ozone and damage forests) are under development.

ALDRIN   Aldrin is a persistent organochlorine insecticide, the use of which is severely curtailed in the developed world. In the UK it caused problems when used as a seed dressing in the 1950s, when birds ate treated seed and passed on residues to predators. In recent years its use has been restricted to controlling insect pests on ornamental flowers such as narcissus, but this has now been banned. The ban was too late, however, to prevent massive contamination of the soil and the river Newlyn in part of Cornwall where aldrin was used extensively. Aldrin is converted to dieldrin in the environment and high levels of this insecticide have been found in fish caught in the Newlyn. One eel contained 220 times the official safe limit of dieldrin for human consumption.

ALGAL BLOOMS   Algal blooms are a consequence of eutrophication (q.v.) – the over-enrichment of waters by plant nutrients. Apart from being unsightly, algal blooms pose a toxic threat to water supplies and to people using affected water, since the algae produce powerful poisons. When they die these toxins are released into the water and are difficult to remove by normal water treatment processes.

In Britain, animals have died through swallowing the algae but there have been no human fatalities. In 1989, eleven soldiers were hospitalised after canoeing on affected water in a Staffordshire lake. In other countries illness such as inflammation of the liver has been reported in people drinking from contaminated water supplies. Some animals can concentrate the toxins in their tissues. High levels were found in shellfish off Britain's east coast in the summer of 1990 and warnings against consuming such foods were issued. In Finland and the Soviet Union deaths have occurred through people eating fish which had concentrated the toxins in their livers.

Algal blooms occur naturally in some areas and not all algae are toxic – it is the blue-green types which produce the poison. Pollution is certainly implicated in many instances and water companies may have to remove phosphate (q.v.) from effluents entering some of the 400 lakes and ponds affected in 1990.

ALUMINIUM   Aluminium is the commonest metal in the earth's crust. It does not normally cause pollution or health problems but there is growing concern that elevated levels of the metal in drinking water may be linked with Alzheimer's disease, a form of senile dementia. This link is based on the finding that patients with Alzheimer's disease may have high levels of aluminium in the brain. Evidence that the metal can harm mental functioning has come from a study of Canadian miners exposed to aluminium powder occupationally. Some 13% of the miners studied showed some mental impairment, compared with 5% in a control group of similar workers not exposed to aluminium, and those miners exposed longest were more likely to be affected.

In parts of the world where rivers and lakes are becoming acidified, levels of aluminium in drinking water are increasing. A correlation between raised concentrations of the metal in drinking water and Alzheimer's disease was found in a study of the problem carried out by the Medical Research Council and the Water Research Centre and published in 1989. Aluminium is used in indigestion mixtures and other medicines, and is present in foods, but it is thought that much of this aluminium is not absorbed. It may be that aluminium in drinking water is more readily taken in by the body and hence presents more of a health risk.

In lakes affected by acid rain, it is the aluminium dissolved from the underlying rocks rather than the acidity which often kills the fish. The metal forms insoluble compounds which block fish gills with a jelly-like material thereby causing suffocation.

There have been four major incidents involving aluminium pollution in Britain, one of which has become a notorious controversy. In July 1988 at the Lowermoor Water Treatment Works, near Camelford in Cornwall, a tanker driver accidentally added 20 tonnes of aluminium sulphate solution to the water supply and some 20,000 people received contaminated water. Immediate symptoms of burning in the mouth and nausea were probably the result of the acidic nature of the compound but local doctors believe that long-term psychological effects, particularly on memory, have occurred in some people. One GP has found symptoms of memory loss and joint pains in 300 people and a neuropsychologist has found signs consistent with a minor brain injury in some of the people affected.

The South West Water Authority, and its privatised successor, disputed the view that long term health effects have occurred, despite the findings of high levels of aluminium in the blood of some patients a year after the incident, claiming that media hysteria was responsible for anxiety in the affected population.

Livestock was also affected by the pollutant. Mental, behavioural and reproductive problems were found in pigs supplied with the contaminated water and many had to be destroyed or died. Post-mortems showed that the animals had accumulated aluminium in the brain and also suffered from other mineral imbalances.

In 1989 a Committee of Inquiry reported that there were no long-term health effects in humans, a view which was rejected by the local community. In the light of subsequent evidence of prolonged damage the inquiry was reopened in 1990. The 'residuary body' of the water authority was convicted of polluting the river with aluminium but the fine was paid by the taxpayer.

Aluminium is often added to drinking water during purification to remove suspended particles such as very tiny bits of peat in water from moorlands, a process which sometimes results in increased levels of the metal in the water. In Britain, at least one million people are thought to drink water containing more than 0.2 mg per litre of aluminium, the EC limit, and there were at least a dozen environmental contamination incidents in the two years following the Camelford problem.

AMMONIA Ammonia is a toxic gas which dissolves in water. It is formed when organic matter containing proteins (or some other compounds of nitrogen) is decomposed by micro-organisms. It is used, chemically combined, in artificial fertilisers. Ammonia is particularly toxic to fish, so when sewage and other effluents are considered for discharge to waterways it is particularly important to minimise their ammonia content. The leachate from domestic waste tips is often high in ammonia and hence must not be allowed to pollute streams.

Ammonia gas is emitted in large quantities from intensive livestock farms; it is a component of the wastes produced by the animals. It is a contributor to acidification since it can be oxidised to nitric acid; this phenomenon appears to be a serious problem in the Netherlands where livestock husbandry is particularly intensive. This is one of the reasons why the Dutch government has placed restrictions on the expansion of such farms.

ASBESTOS   Asbestos is a naturally occurring material – a mineral – which exists in three main forms. These are chrysotile (white asbestos), amosite (brown asbestos) and crocidolite (blue asbestos). Until the late 1960s, when its health risks were better appreciated, asbestos was used widely for its excellent insulating properties against heat. Buildings, power stations and steam engines contained large amounts of the material and smaller quantities were used in or on domestic appliances such as ovens and ironing boards. Asbestos sheets have been used for roofing and as cheap wall panels while the fibres have been used to reinforce cement in tiles and pipes.

Although some concern has been expressed about risks from swallowing asbestos in drinking water, the main hazard from the substance derives from inhaling it. The fine fibres which make up the material can penetrate deep into the lung and set up lung cancer or a much rarer cancer of the chest wall called mesothelioma. Smokers exposed to asbestos find that their chances of getting lung cancer are multiplied; a smoker is eleven times as likely to contract cancer as a non-smoker, a non-smoker exposed to asbestos is five times as likely to contract lung cancer as someone who is not exposed, while a smoker exposed to the same level of asbestos is fifty-three times as likely to contract lung cancer as a non-smoking non-exposed individual.

Asbestos-induced lung cancer and mesothelioma take a long time to appear and can result from a comparatively low exposure. Very heavy exposure to asbestos leads to another condition called asbestosis. Here the lungs clog up as they react to the presence of large numbers of fibres and the victim finds breathing extremely difficult; physical exertion becomes impossible. Asbestosis is an industrial disease which does not usually occur in the developed world today. Asbestos miners and processors in the Third World may still contract it, however, as hygiene standards there are often much lower.

It was once though that blue asbestos and, to a lesser extent, brown were much more dangerous than white asbestos but it is now realised that it is prudent to regard all forms as equally dangerous. The blue and brown forms are now banned in Britain but some white asbestos is still used. Much remains in buildings, in wall panels, roofing materials and even some textured paints. Such products do not present a hazard to health unless abraded, as by sanding or drilling, and are best left in place unless they begin to deteriorate. Asbestos removal is a highly skilled and tightly regulated activity not to be undertaken by amateurs. Once removed, the most suitable fate for asbestos is in a deep landfill as it is not practical to destroy it under normal circumstances. If it is not buried sufficiently deeply it may be mobilised when the land is redeveloped – no-one knows how many pockets of former industrial land remain contaminated by asbestos.

BENZENE Benzene is a toxic and carcinogenic hydrocarbon which is known to cause leukaemia. It is present in petrol with some unleaded versions containing more than corresponding leaded fuels. It causes air pollution around petrol stations as vehicles are fuelled and also escapes as an unburnt hydrocarbon from vehicle exhausts and petrol tanks.

See also Motor vehicles.

BERYLLIUM Beryllium is a light metal which is used in some specialist alloys, in nuclear reactors and in nuclear weapons. It is not radioactive but is highly toxic, causing damage to the lungs which may prove fatal. Beryllium and its compounds are also carcinogenic. In the autumn of 1990, a large escape of beryllium followed an explosion and fire at a Soviet nuclear fuel plant. The homes of at least 120,000 people were contaminated and levels of beryllium in parts of the centre of the affected city, Ust-Kamenogorsk, reached 890 times the permitted limit. The environmental and health consequences of this accident will not become clear for some years. Little is known about the behaviour of large escapes of this material in ecosystems and health monitoring of exposed individuals has been scanty.

BHOPAL One of the world's worst pollution disasters occurred in December 1984 when water entered a chemical storage tank at the Union Carbide plant at Bhopal, India. A vigorous reaction occurred between the water and methyl isocyanate, the chemical in store, and a massive cloud of gas was ejected from the tank. There is some dispute as to what the cloud contained. Certainly it was mostly methyl isocyanate, but it may also have contained hydrogen cyanide and the war gas phosgene: in any event it was highly toxic. As a result of exposure to the gas, which formed a dense layer at ground level, some 2,500 people died and many thousands suffered serious damage to their eyes and lungs. Rates of spontaneous abortions and still-births increased dramatically and the total number of people affected to one degree or another is estimated at 200,000.

Over four years later, in February 1989, those harmed or bereaved by the accident were awarded $470 million in compensation in an Indian Supreme Court judgement against the plant operators – a figure which failed to satisfy those affected.

BIOACCUMULATION   Bioaccumulation is the process by which persistent pollutants such as heavy metals are concentrated in the tissues of living organisms. Many shellfish, for instance, feed by filtering their food from water which they pass through their shells. Particles of heavy metals contained in the water are retained and metal levels rise – which is why 'mussel watch' programmes can be used to monitor metal pollution around coasts. Organochlorine pesticides such as DDT and dieldrin also bioaccumulate because they are readily soluble in fat. Bioaccumulation enables toxic materials to be passed up food chains since predators feeding on small animals which have absorbed persistent materials will themselves accumulate the substance in question. Bioaccumulation of heavy metals has resulted in limits or bans being set on the consumption of fish and shellfish, while pesticide bioaccumulation has had serious effects on the fertility of such species as seals, otters and birds of prey.

BIOCONCENTRATION   see Bioaccumulation.

BIODEGRADATION   Strictly speaking, biodegradation means the breakdown of a substance or material in the environment by the action of living organisms; but the term is used more generally to include breakdown by weather, air and other outdoor factors. Living things, usually micro-organisms, are of primary importance in converting complex materials into simpler ones by using them as food. The materials in question could be as bulky as discarded paper or as diffuse as detergent residues in a river. Some plastics are now marketed as being biodegradable although in some instances only a proportion of the material is actually destroyed.

Biodegradable materials do not normally accumulate in food chains and disappear from the environment fairly quickly. The presence of toxic materials, however, can inhibit biodegradation by harming the micro-organisms responsible. Most naturally occurring chemicals are biodegradable and the majority of synthetic compounds are as well. Some artificial substances, however, do not resemble compounds in nature and are hence not easily used as food by living things. This means that they can persist for long periods and may bioaccumulate. Elements, such as heavy metals, cannot be biodegraded as they are not convertible into simpler forms.

BIOLOGICAL CONTROL
Biological control is the process whereby living organisms, instead of chemicals, are used to control pests. The oldest example is keeping a cat to kill mice, but much more sophisticated techniques now exist for the control of agricultural pests. Most biological control methods involve predators or parasites of the pest. These are bred in large numbers and released into the affected area. Work has also been carried out on diseases of pests and a toxin derived from a bacterium that infects insects has been in use for many years.

Moves to use genetically engineered micro-organisms are under consideration but there is considerable concern about the safety of releasing 'artificial organisms' into the general environment. A species of wasp has been used successfully to control pests in greenhouses and this is commercially available. Biological control will not completely eliminate the target pest (and neither will chemical methods) but it can reduce levels of infestation to a tolerable level without the risk of pesticide residues or the accidental poisoning of other species.

BIOLOGICAL TREATMENT
Effluents and wastes containing organic matter can often be treated biologically. Indeed this is the whole basis on which a sewage works operates. Micro-organisms such as bacteria and protozoa use organic materials as food and, providing that the effluent does not contain substances toxic to them, they can break down much of the noxious material in the effluent, leaving simpler and generally less polluting materials behind.

A traditional sewage works involves trickling sewage onto a biological filter which consists of a bed of coke or similar material which has large populations of micro-organisms on the surfaces of the solid. As the effluent passes through, waste materials are removed by aerobic (oxygen-using) organisms and some of the harmful bacteria are consumed. The BOD of the effluent (see below) is thus greatly reduced. The resulting liquid may need further treatment to remove dissolved materials such as nitrates and this can also be done biologically.

More modern treatment plants use the activated sludge process, where oxygen is pumped into a continuously stirred tank containing the microbes. Anaerobic organisms (those which do not use oxygen) can be used to digest sewage sludge to produce methane which can then be used as a fuel. A wide range of wastes, including some hazardous materials, can be treated biologically and the principle is also being extended to cleaning up contaminated land.

BOD   BOD, which is short for Biochemical Oxygen Demand, is a measure of how polluting an effluent is likely to be if discharged into a stream. Any effluent containing organic matter provides food for micro-organisms which occur naturally in water. As they use this food they respire and use up the oxygen dissolved in the water. If they consume oxygen faster than it is replaced from the air or by plants, the stream becomes depleted in the gas and may become malodorous. The range of creatures which it can support becomes restricted. Effluents with a high BOD, such as raw sewage, silage liquors and waste from food processing can cause serious water pollution if not treated before discharge, especially where they are not well diluted by the river. The aim of many effluent treatment plants is to reduce the BOD of the waste before discharge.

CADMIUM   Cadmium is a highly toxic heavy metal which has caused serious environmental pollution in many places. It often occurs in nature together with zinc and lead, so smelters processing these ores often release cadmium as well as the other metals. Like other heavy metals, cadmium is persistent in the environment and can accumulate in living organisms. Very high levels of the substance have been found in shellfish living near a smelter and old mine workings in the Bristol Channel.

Cadmium has a range of industrial uses. It is used in plating metals, batteries, pigments and stabilisers in plastics and various alloys. Because of problems caused by cadmium pollution, the EC is to impose restrictions on the uses of cadmium in products where suitable substitutes exist.

The metal enters the environment in a number of ways. Effluents from metal plating works and the run-off from scrap yards contain cadmium while the incineration of domestic waste releases the material into the atmosphere. The production of phosphates, for detergents and fertilisers, from phosphate rock results in the discharge of cadmium-containing effluents to the sea, the amounts involved depending on the source of the phosphate rock. Metal-bearing wastes are another source of cadmium; the Somerset village of Shipham is heavily contaminated in places by cadmium derived from old zinc mines while some of the cadmium entering the Severn estuary derives from mining wastes and natural mineralisation. Sewage sludge often contains elevated levels of cadmium and when deposited on the land leads to soil contamination. This, in turn, can contaminate growing crops and appear in the meat, especially offal, of animals grazing on affected land. The major source of cadmium for most people is diet, but smokers derive a relatively large amount from tobacco.

Like lead, cadmium has no function in the human body. It can accumulate in the liver and kidneys and may cause kidney disease in the long term. Chronic exposure may also cause heart disease although the levels required to do this are uncertain. Prolonged exposure to high levels of cadmium in the diet can result in a disease known as itai itai – Japanese for 'it hurts, it hurts'. This disease killed about 100 people in Japan in the 1960s as a result of the consumption of rice grown on soil contaminated with cadmium-bearing wastes. Degeneration of the bones and multiple fractures were the main features of the disease. The inhalation of large amounts of cadmium can be fatal; welders breathing in fumes from heated cadmium-plated metal have died as a result. Cadmium is also carcinogenic.

The measures proposed by the EC to reduce cadmium usage will help to restrict the amounts of cadmium entering the environment but other trends may act against them. Rechargeable nickel-cadmium batteries are becoming increasingly popular and although they last a long time they are eventually discarded in wastes which may be incinerated. Few such batteries are recycled. As the ban on dumping sewage sludge at sea comes into force, more cadmium may end up on the land unless measures are taken to reduce the amount of the material entering the sewerage system in the first place. The hazards of lead have been investigated exhaustively over the past two decades, but the spotlight may now focus on cadmium.

CAMELFORD   see Aluminium.

CARBON DIOXIDE   Carbon dioxide is a colourless gas formed when any material containing carbon burns in air. It is not toxic, although it will suffocate animals breathing very high concentrations, and is a natural component of the atmosphere. Carbon dioxide is extracted from the atmosphere by plants and used to make sugars in the process called photosynthesis, the base of most food chains. It is also released when most living organisms respire.

Levels of carbon dioxide in the air were much higher in the geological past but the gas extracted by growing plants was removed from circulation when the plant remains were buried in sediments. These remains then formed fossil fuels and when they are burned today they return carbon dioxide to the atmosphere.

Levels are now rising: the increase was slow in the early part of this century but between 1958 and 1988 the concentration of carbon dioxide in the air at the monitoring station at Mauna Loa rose from an average of 315 ppm to nearly 350 ppm. The rate of increase is growing as fossil fuels are used up at an increasing rate and as rainforests are burned and not replaced. This is of great concern since carbon dioxide is a greenhouse gas (see Greenhouse effect).

The growth of plants is not the only mechanism by which carbon dioxide is removed from the atmosphere. It dissolves in the oceans and is used by marine creatures to make calcium carbonate for their shells. These shells may end up in sediments and be converted into limestone or other, similar, rocks.

CARBON MONOXIDE   Carbon monoxide is a toxic gas which is produced when carbon-based fuels such as oil, gas and coal are burned with insufficient oxygen present. It is emitted by motor vehicles in large quantities and also by defective heating appliances such as gas boilers with blocked flues.

The gas is poisonous because it combines with the red pigment, haemoglobin, in the blood. This substance normally carries oxygen from the lungs to the tissues, but carbon monoxide blocks the uptake of oxygen. Low-level carbon monoxide poisoning results in headaches, dizziness and sleepiness, while breathing air containing high levels of carbon monoxide is rapidly fatal. Low-level exposure can also damage the heart and aggravate heart disease; some of the ill effects of tobacco on the heart are attributed to carbon monoxide in the smoke. The reduction in birth weight of babies born to smoking mothers is also ascribed to carbon monoxide.

Levels of carbon monoxide are high in heavily trafficked areas, especially where vehicles are moving slowly and dispersion is inhibited by buildings. Car parks and other enclosed spaces are also high risk areas. Major health effects on the average motorist have not been proven but some sectors of the population, particularly those with heart disease – and especially those who smoke as well – may be at risk.

Carbon monoxide emissions can be controlled by ensuring adequate levels of oxygen when fuels are burned and making sure that combustion gases from heating appliances do not leak into homes. Rooms heated by paraffin or bottled gas should be well ventilated. Emissions from motor vehicles can be reduced by ensuring that the engine is properly tuned and by fitting a catalytic converter.

CARS   see Motor vehicles.

CATALYTIC CONVERTER A catalytic converter is a device which is fitted into a motor vehicle's exhaust system to remove polluting gases. The technology was developed in the USA in response to the photochemical smogs which afflict Los Angeles but it is now being used in Japan and Europe. Few cars sold in Britain are fitted with catalytic converters at the moment but changing European standards for exhaust emissions are set to alter this.

There are several designs of catalytic converter but two main types are in use. All types aim to destroy carbon monoxide and unburnt hydrocarbons as well as reducing emissions of nitrogen oxides. The simple oxidation catalyst burns off carbon monoxide and hydrocarbons, achieving reductions in nitrogen oxides by recirculating some of the exhaust gases into the engine. The three-way catalyst also burns hydrocarbons and carbon monoxide but, in addition, converts nitrogen oxides back to nitrogen which is a normal constituent of the air.

Catalysts, especially the three-way version, are effective at controlling vehicle emissions but have some disadvantages. They add to the vehicle's fuel consumption – a 9% fuel penalty has been found over a complete test cycle although this drops to about 3% at motorway speeds – but good design can alleviate this. They also require the use of precious and rare metals such as platinum. Furthermore they are 'poisoned' by lead if leaded petrol is used in them, so vehicles fitted with catalysts should only use unleaded fuel. Another drawback is that in order to function efficiently they need to operate at high temperatures which are only reached after the engine has been running for some time. Thus for short journeys of less than a kilometre, catalytic converters achieve little in the way of emission controls.

CFCs CFC stands for chlorofluorocarbon, a hydrocarbon in which the hydrogen has been replaced by chlorine and fluorine. Chlorofluorocarbons are particularly useful because they are inert – they do not react chemically with many other substances except under extreme conditions. They are valuable in aerosol cans, fire extinguishers and in blown insulating foams. CFCs are also used in refrigerators and freezers to transfer heat out of the cold compartment.

When released into the air, as all CFCs eventually are, these materials drift upwards to the stratosphere. Their very inertness means that they are not broken down to any great extent in the lower parts of the atmosphere. High up, they are exposed to intense ultraviolet light (q.v.); this splits up the molecule, producing chlorine atoms. These are highly reactive and will combine with many other substances: one of the materials for which they have a particular affinity is ozone (q.v.). The attack on an ozone molecule starts a chain reaction which regenerates the chlorine atom. This then goes on to attack some more ozone. One chlorine atom, from one CFC molecule, can destroy thousands of ozone molecules before it combines with something else and becomes harmless. Thus small quantities of CFCs can do a lot of damage to the protective ozone layer.

CFCs are also powerful greenhouse gases, some being more than 7,000 times as effective at trapping heat as carbon dioxide.

Since the discovery of a large hole in the Antarctic ozone layer, and strong evidence linking this hole with the presence of CFC breakdown products, international moves to restrict the use of these compounds have progressed. Pressure from Friends of the Earth quickly led to the removal of CFCs from most aerosol products in the UK and some manufacturers of refrigeration equipment are reducing, or removing completely, CFCs in insulating foams. At the time of writing, acceptable substitutes for CFCs in all their applications – particularly the most popular compounds known as CFC11 and CFC12 – are not commercially available, although the major CFC manufacturers are all racing to produce safe and effective products by the most economical routes.

The favourite compounds under investigation are known as HFCs and HCFCs – fluorocarbons and chlorofluorocarbons which contain a hydrogen atom in their molecules. This makes them much more unstable than CFCs and they will tend to break down in the lower atmosphere so that any chlorine in the molecule (there is none in an HFC) does not reach the stratosphere in significant quantities. There are some problems, however. Some of the materials are flammable and may pose a hazard if used in refrigerators. They may be converted in the environment to the highly toxic and persistent material trifluoroacetic acid, although this has not been proven. The HCFCs will still have a small ozone depleting effect but, more importantly, HFCs and HCFCs are powerful greenhouse gases, about as effective as CFCs at trapping heat.

Current proposals are for a complete phase-out of CFCs and halons (q.v.) by 2000, except for a few essential safety uses. Carbon tetrachloride, an ozone depleting solvent, is to be phased out by 2000 with an interim reduction of 85% by 1995. HCFCs are to be used carefully and phased out by 2040–2060. Countries such as China and India, where refrigerators have hitherto been owned by only a few people, are demanding the right to use the gases or be provided with acceptable substitutes as their citizens come to expect a higher standard of living. Under the agreement to restrict these substances signed in London in 1990, they are to be given an additional ten years to achieve the above targets.

Because of their long lifetimes in the atmosphere, and the long time it takes for CFCs released at ground level to reach the stratosphere, we do not know how much damage CFCs already released will do to the ozone layer. Even if the use of all of them were banned tomorrow, it could be fifty years before the total cost of their use is realised.

CHERNOBYL On 26 April 1986 a major accident occurred at the Chernobyl nuclear power station complex near Kiev in Russia. An explosion blew clouds of radioactive material into the atmosphere and this was carried by winds over much of northern Europe, falling out onto the land as it went. The result was widespread contamination of the food chain with the radioisotope caesium-137, a material which is accumulated by some plants, notably mushrooms, and difficult to remove from soil.

The accident occurred because the plant operators, who were carrying out an experiment, turned down the power level in one of the Chernobyl reactors to below the minimum safety level – these reactors have to operate at above this level otherwise runaway reactions may occur. They also shut down the system which cools the nuclear core in an emergency as well as other safety systems. When a runaway reaction did occur the reactor could not be shut down and part of the fuel exploded. This was followed by a hydrogen explosion which blew the lid off the reactor, spewing a massive cloud of radioactive material into the sky.

There were no early warnings of the accident and most countries affected had to rely on their own radiation monitoring systems. One of the first alarms was raised in Sweden where a monitor at the Forsmark nuclear power station detected increased radiation levels. In Britain, some of the most useful initial measurements were carried out at a school rather than by official bodies. There was a widespread lack of information about what was happening and what the consequences might be; as a result, advice on how to reduce the risk of contamination was not forthcoming in many countries. In Britain, ministers averred that there was no real problem without the benefit of evidence.

Across northern Europe crops and livestock were contaminated. The Lapps were probably the worst affected group of people; lichens on which reindeer feed readily accumulated the radioactive material and the livestock soon absorbed large amounts. Thousands of reindeer had to be slaughtered and dumped, wreaking havoc on the food supply and lifestyle of the Lapps. Some countries issued iodine supplements to children and pregnant mothers, in order to block the uptake of a radioisotope of iodine, while others advised against the consumption of rainwater. In Britain the worst contamination was in the sheep-farming uplands of Cumbria and North Wales, although subsequent work showed pockets of high radioactivity in other areas. Even five years later, some Welsh sheep farmers were unable to sell their stock because it was too radioactive. The caesium-137 is constantly recycled between the thin soil, grass, sheep and their droppings.

Much of the area around Chernobyl remains badly contaminated and the reactor itself is sealed in concrete. The incident had a profound effect on public attitudes to nuclear power and brought home to many people the difficulties involved in containing and monitoring the effects of a major nuclear accident.

CHLORACNE   Chloracne is a painful skin condition caused by exposure to PCBs (q.v.) and related compounds. It is not certain whether the condition is caused by pure PCBs or by the dibenzofurans and dioxins which contaminate them; certainly the contaminants alone can cause the disease. One of the main consequences of the Seveso incident (q.v.), in which dioxins were released, was an outbreak of chloracne in the surrounding population.

CHLORINATED HYDROCARBONS   A chlorinated hydrocarbon is a chemical which consists of the elements chlorine, hydrogen and carbon. There are many such compounds and they range from fairly simple solvents such as dichloromethane (used in paint stripper) to complex pesticides such as dieldrin. Chlorinated hydrocarbons are generally stable in the environment although they can be broken down by the action of ultra-violet light (see CFCs). They are not readily biodegraded since such materials are rare in nature and micro-organisms do not naturally use them as food. They are more soluble in fat than in water and this is why they can be bioaccumulated, although the more volatile simpler types do not remain in body tissues for long. Many are toxic and some are carcinogenic. The chlorinated hydrocarbon insecticides such as DDT and dieldrin have caused serious harm to wildlife in many parts of the world.

CHLORINATION
Chlorination simply means 'adding chlorine to'. It is used by chemists to mean combining chlorine with another molecule – e.g. to make polychlorinated biphenyls – and by water supply engineers to mean disinfection of water using chlorine.

Water for domestic consumption is filtered and chemically treated to remove various undesirable substances such as suspended solids and colours. It may still harbour harmful bacteria, however, so it is disinfected before being put into public supplies. Chlorine is the main substance used since it is effective against most micro-organisms and also leaves residues in the water which can help to overcome any subsequent contamination by bacteria.

There are some problems with chlorination, however. Excessive amounts can leave an unpleasant taste but, more importantly, if the water is already contaminated with organic matter the chlorine may react with some of the contaminants. This can lead to stronger tastes (like the antiseptic TCP) and may also form toxic and mutagenic compounds such as trihalomethanes. Many water supplies contain traces of organic matter and experiments have shown that such water is more prone to cause mutations in bacteria after chlorination than before. The significance of this phenomenon for human health is unclear but it may be cause for concern since mutagenic chemicals often cause cancer.

The use of chlorine for the disinfection of swimming pools has also caused problems. Apart from accidents with the chemicals used to produce the material, which often lead to people being hospitalised with chlorine poisoning, the chlorine produces potentially harmful chemicals on contact with substances in urine and other bodily secretions. Some of these chemicals are irritant and cause sore eyes.

CHLORINE
Chlorine is a greenish highly poisonous gas which was used against troops in the First World War. It is a major industrial chemical with many and varied applications. Chlorine is a very reactive element – it combines easily with other substances – but some of its compounds are quite resistant to attack by other materials, examples being halons (q.v.), CFCs and some solvents. The storage of large amounts of chlorine gas can present an environmental hazard – there have been a number of accidental releases of the substance from manufacturing plants and a major leak could cause many deaths if it drifted over a populated area. Accidents involving chlorine have happened in the home when acid toilet cleaners have been mixed with bleach, the result being a sudden release of the gas.

Many brands of domestic and industrial bleaches are based on chlorine. They are used for a variety of purposes including bleaching fabrics and paper as well as for disinfection. When these bleaches are used small quantities of dioxins and dibenzofurans (q.v.) are formed and these may be released to the environment and/or retained in the product. Traces have been found in paper, disposable nappies and even milk cartons as well as in the effluents from paper mills. The significance of these small quantities of dioxins and dibenzofurans is uncertain, but in order to remove any risk of exposure to them some manufacturers are switching to alternative bleaches which do not produce these toxic materials. Many paper and pulp based products now carry the message 'non-chlorine bleached'. Chlorine-free bleaching is also accepted as part of the definition of 'environment-friendly paper'.

CONTAMINATED LAND

The use of land for waste tipping, chemicals manufacture, gasworks and many other activities has left many areas contaminated and unsuitable for use for normal purposes. The presence of hazardous chemicals in the ground poses a risk to health and may also damage foundations, pipework and cables. In some instances, contaminated land may be obvious: plants fail to grow, the ground may smell or noxious liquids may appear on the surface. In other cases the risk may be more insidious: toxic metals, surface deposits of asbestos and some organic pollutants may not be detected unless sophisticated chemical analyses are carried out.

The extent of contaminated land in Britain is unknown and it is only now that proposals for a contaminated land register are being given serious consideration. In some areas with a history of industrialisation, contamination is likely to be widespread. A pilot survey in Cheshire revealed 1,577 potentially contaminated sites. Occasionally incidents come to light such as the case of a woman who bought a piece of land only to find out subsequently that it was the site of an old waste tip now leaking PCBs – and that she was legally responsible for the enormous cost of dealing with the problem.

Cleaning up contaminated land is no easy task. In some instances, biological treatment can be used to destroy organic contaminants such as polycyclic aromatic hydrocarbons. Expensive techniques for extracting solvents have been developed in the USA and it is also possible to solidify the soil into a glassy material. In some instances it may be necessary to take away the polluted material and dump it in a controlled landfill, as happened in Bristol where a city farm found that its land was contaminated by metals from an old smelting works.

CRYPTOSPORIDOSIS Cryptosporidosis is a disease caused by infection with the micro-organism *Cryptosporidium*, a parasite present in animals and their wastes. Humans are not often affected – about 2% of the cases of diarrhoea investigated are caused by this organism – and most cases result from direct contact with animals. Several outbreaks have occurred recently as a result of drinking water contamination, notably a series of cases in Swindon and part of Oxfordshire in 1989. The cause was heavy pollution of a reservoir with animal wastes. The water treatment processes were unable to remove all the parasites, so contaminated water entered the public supply.

The disease is not usually fatal although it is unpleasant, the symptoms being nausea, vomiting and diarrhoea. Small children are more likely to be badly affected and anyone whose resistance to infection is lowered may suffer lasting disability.

Cryptosporidium is difficult to detect so water supplies are not monitored routinely for its presence but water companies have been advised by an expert group which looked at the problem in 1990 to draw up monitoring strategies which could be used if necessary. The development of emergency control measures is also recommended.

DDE see DDT.

DDT An organochlorine compound, DDT was the first cheap synthetic insecticide to be produced on a wide scale. It was first deployed in the 1940s and undoubtedly saved many lives by controlling disease-carrying insects at the end of the Second World War. It was also used widely as an insecticide on crops in many countries.

Unfortunately the miracle turned sour. DDT was found to be highly persistent in the environment and residues were passed up food chains, affecting species at the top. The material is very soluble in fats which means that it is readily accumulated in the tissues of animals. DDT is not highly toxic in the acute sense but it – or rather its breakdown product DDE – affects the metabolism of birds of prey so that they produce thinner eggshells which crack when the birds sit on them. DDT was one of the chemicals responsible for the decline of British birds of prey such as the peregrine in the 1950s and early 1960s. Humans, being at the top of the food chain, also receive DDT and at one point the milk from some American women contained so much of the chemical that it would have been illegal to sell it if it had been cows' milk. DDT is regarded as carcinogenic by some authorities.

Restrictions on DDT use have been in force in developed countries for some time and it is banned in the EC. It is still used in public health and agricultural programmes in developing countries, although many pests are now resistant to it. Food imported from these countries frequently contains DDT residues and migratory birds bring the material back to Europe after visiting Africa, where it continues to deplete populations of birds of prey. Such is the persistence of DDT that traces of it have been found virtually everywhere although in many instances this is of little significance in health terms.

DETERGENTS In the 1950s and early 1960s some rivers were plagued by detergent 'swans' – huge masses of foam produced when water containing detergent residues ran over weirs or other sources of turbulence. Unlike traditional soaps, these detergents were not readily biodegradable and hence passed through sewage works without being completely destroyed. The results were foaming, reduced aeration of water and, sometimes, toxicity to fish and other water life.

The detergents in question were 'hard' detergents which have been replaced in all domestic washing powders and liquids by soft detergents which biodegrade readily. This was a voluntary move by manufacturers, although legislation undoubtedly would have been introduced had this not happened.

Modern washing powders may present another problem to the aquatic environment as a result of their phosphate content. Phosphates (q.v.) are plant nutrients and can cause eutrophication (q.v.) in waterways. Some 20%–25% of the phosphates discharged to rivers and lakes come from detergents although this percentage will vary from place to place according to the magnitude of other sources. Eutrophication seems to be a growing problem and there are undoubtedly some areas where phosphates from detergents tip the balance between a healthy and an over-rich system.

There is a wide range of 'environment-friendly' detergent products on the market. These are phosphate-free and all biodegradable; some claim to biodegrade quicker or more completely than conventional products. They also do not contain a range of other materials such as optical brighteners which make white clothes look bluish (associated with cleanliness) and which may leave residues. Some of them are derived from vegetable oils, with a certain amount of chemical processing, instead of being made wholly from petroleum, in an attempt to conserve non-renewable resources.

DIBENZOFURANS see Dioxins.

DIELDRIN Like DDT, dieldrin is an organochlorine insecticide but it is much more toxic than the former compound. It has been used for pest control on crops, in public health programmes and in buildings for the control of wood-boring insects. Dieldrin is persistent and a carcinogen, although British authorities have not accepted the cancer data which led the US authorities to ban it. The material has caused serious harm to wildlife around the world and in Britain has contributed to the decline of birds of prey. It has also been responsible for reducing the populations of British otters by interfering with their fertility. High concentrations of dieldrin have been found in eels caught in several British rivers, notably the Newlyn in Cornwall which received run-off from fields growing ornamental flowers. Aldrin, used against pests on the flowers, was converted into dieldrin in the environment and high levels persisted in the soil and river: one eel contained 220 times the official safe limit for dieldrin in food.

DIESEL Diesel fuel is used in vehicle engines which do not need spark plugs to ignite the fuel. It is less volatile (and hence less flammable) than petrol and a correctly adjusted diesel engine can be fairly clean. Many diesel engined vehicles are not properly maintained, however, and emissions of pollutants, notably smoke, are often high. Carcinogenic materials are also emitted from diesel engines – see Polycyclic aromatic hydrocarbons. Diesel fuel is never leaded.

See also Motor vehicles.

DIOXINS AND DIBENZOFURANS When chlorinated compounds are burned at moderate temperatures and oxygen is not plentiful, destruction may be incomplete and a wide range of partial combustion products may be formed. They often include the compounds known as chlorinated dioxins and chlorinated dibenzofurans (the word 'chlorinated' is often dropped for convenience). These substances are persistent in the environment and some are highly toxic.

Dioxins and dibenzofurans may be formed when chlorine-based bleaches are used to whiten paper and wood pulp, so traces have been found in a wide range of such products as well as in effluents from plants producing them. Waste incinerators and crematoria produce measurable quantities of dioxins – the former from chlorinated materials in the waste stream and the latter from plastic coffins. Fires involving electrical equipment containing PCBs cause the formation of dibenzofurans in particular. One of the world's most notorious industrial accidents, at Seveso (q.v.) near Milan in 1976, released a large quantity of dioxin into the atmosphere and caused widespread land contamination.

There are 75 types of chlorinated dioxin and 135 chlorinated dibenzofurans. While some have been investigated thoroughly, little is known about the toxicity of many of these forms and even less about the combined effects of various mixtures. One of the most toxic dioxins, TCDD, is known to be highly poisonous to mammals (guinea pigs are unusually sensitive) and to cause cancer and birth defects in laboratory studies. In humans these materials can cause a painful and unsightly skin disease called chloracne (q.v.). Exposure to TCDD present in defoliants is thought to have caused birth defects in Vietnamese people sprayed with them during the Vietnam war. There have been few, if any, documented cases of humans dying as a result of exposure to these compounds but horses and other non-human animals died when roads and a horse arena in Kentucky were sprayed with oil, contaminated with dioxin, as a dust suppressant.

Traces of dioxin and dibenzofurans are present nearly everywhere and they originate almost entirely from human activities. They bioaccumulate and are known to contaminate foodstuffs and both human and cows' milk. The importance of widespread dioxin contamination, in health terms, is uncertain, but since the toxicology of the many materials involved has not been investigated fully, some authorities are concerned that a significant health problem exists.

DRINS The 'drins' is a group of persistent organochlorine insecticides comprising aldrin, dieldrin and endrin. All are toxic to humans and all have caused serious environmental problems in various parts of the world. Aldrin and dieldrin have been used more extensively in the UK and are discussed separately. Endrin has not been so popular and its use is now banned; a gamekeeper was fined in 1990 for using it to kill predators on a Scottish estate.

DRINKING WATER Drinking water in the UK is generally regarded as being pure and wholesome, certainly when compared with that available in some other parts of the world. There are, however, problems of contamination and some of these have put Britain in breach of EC law.

Two types of contamination affect water: microbiological and chemical. Microbiological contamination of British water is rare, although there have been problems in some areas with the organism *Cryptosporidium* which can cause serious intestinal and stomach problems in people consuming it. *Cryptosporidium* is very difficult to kill by normal purification methods and it is a growing problem (see Cryptosporidosis).

Chemical contamination is more widespread. Metals, nitrates and pesticides all find their way into drinking water supplies in one place or another and there is also concern about solvent residues.

The most serious metal contaminant, apart from accidental incidents such as the Cornish aluminium problem (see Aluminium), is lead (q.v.). For many years lead was used for pipework and, although this practice has been discontinued, many properties still contain lead plumbing. Small amounts of lead are dissolved as the water moves through the pipes and even more goes into solution when water remains still in the pipes overnight.

Soft waters dissolve lead more easily than hard; hence some water companies add hardening chemicals to the water to prevent the dissolution of lead. In some instances particles of lead may be abraded from pipe walls; levels of up to 1,000 times the legal limit were found in water in some homes in Blackburn following incidents when dirty water full of suspended material came through the taps.

In England in 1976 some 7% of homes were estimated to exceed the EC limit for lead in drinking water of 100 microgrammes of lead per litre. In Scotland the figure was 34%. Since that survey some remedial work has been carried out and the limit has been reduced to 50 microgrammes per litre by British law. Many thousands of houses must be over this limit and the UK programme for dealing with the problem in areas of particular concern is not expected to be complete for many years. Householders should certainly consider removing lead pipework where this is possible: grants have been available from local councils to assist with the costs of so doing.

Aluminium (q.v.) is another metal which sometimes contaminates drinking water, usually through its over zealous use in water treatment. Aluminium compounds are added to water during purification to remove very small particles of solid matter suspended in the water and should precipitate out. Some residual aluminium remains and this may reach levels which cause concern. About a million people in Britain are thought to drink water containing more aluminium than the EC limit.

A third inorganic contaminant found in water is nitrate (q.v.), which has been found at levels in excess of EC and World Health Organisation standards. In 1976 bottled water was distributed to the mothers of young babies in parts of southern England because the tap water contained too much nitrate for it to be usable in making up bottle feeds. Nitrates contaminate rivers and groundwater as a result of sewage discharges and agricultural fertiliser use.

Organic chemicals are also present in some drinking waters and these may be natural in origin, as when water drains off peat moors, or present because of human activities. Synthetic substances include pesticides, solvents and some drug residues. Some of this contamination results from the re-use of water; sewage treatment processes and drinking water purification plants do not always remove every trace of these chemicals. In other instances, contamination may result from activities above groundwater or in rivers from which water is abstracted. A leak of phenol-containing industrial effluent into the River Dee led to the supply of foul-tasting polluted drinking water to some households, while solvents carelessly discharged on the surface have contaminated groundwater in Hampshire, Oxfordshire and Suffolk. The chlorination (q.v.) of water prior to supply can lead to the presence of organochlorine compounds in drinking water and this may be a cause for concern.

Pesticide use also leads to water contamination either by run-off into streams or by percolation through the soil to groundwater. Persistent herbicides such as simazine and atrazine, used on farms, railway tracks and local authority property, have contaminated many aquifers. Insecticides such as lindane have also been detected in drinking water. A survey carried out by Friends of the Earth in 1987 showed that 298 English water supplies exceeded the EC limit for a single pesticide, while 76 sources breached the limit for total pesticides. (The Maximum Admissible Concentration (MAC) for any one pesticide is 0.1 microgramme per litre; for total pesticides the MAC is 0.5 microgrammes per litre). Friends of the Earth believe that these results seriously underestimate the problem and that water suppliers do not test their product for pesticide residues frequently or extensively enough. The British Medical Association has also expressed its concern about pesticide contamination of drinking water.

ELECTROMAGNETIC RADIATION   Many of the electronic devices which we use emit electromagnetic radiation as they operate. In addition, we are bathed in this radiation in the form of radio, television and other communications transmissions. People living near power lines are also subjected to electromagnetic fields as electricity is carried along the wires.

There are many different types of electromagnetic radiation including light, heat, radar waves, microwaves and X-rays. Each type has a range of frequencies and wavelengths (see Chapter 1) and most do not present a hazard to health unless exposure is so massive as to cause burns. However, some people are becoming increasingly worried that the types of radiation emitted by electronics, power lines etc. may be harmful and in a few instances health problems have been linked with very close proximity to power lines.

The evidence for more widespread risks is very tenuous at the moment but little research has been carried out. The radiation involved is different from the ionising radiation resulting from radioactivity (q.v.) and it remains an open question as to whether harm is being caused and, if so, how much. Given that electromagnetic radiation of these types has been with us for a long time it seems improbable that a massive problem exists. However, levels of exposure are rising and new frequencies are becoming used, so complacency may not be justified.

ELECTROSTATIC PRECIPITATOR
This device is used to remove smoke and other particles from effluent gases passing up chimneys. It consists of two electrodes – plates or a wire running through a pipe – with a strong electric field applied across them. As the particles pass between the electrodes they become electrically charged and stick to one of the electrodes, in much the same way that dust sticks to the front of a television screen. The collected particles are removed and disposed of appropriately and the cleaned gases are released to the atmosphere. Electrostatic precipitators are very efficient at removing particles but they do not remove gases: for these a scrubber (q.v.) of some sort is necessary.

EUTROPHICATION
Eutrophication is the process which occurs when a waterway becomes too rich in plant nutrients. Nitrates, phosphates and some other, minor, substances are needed by plants for growth and in an undisturbed lake or river these will be recycled between the water, muds and living things. Over many years a lake will become richer and then over-rich (eutrophic) but human activities can accelerate this process manyfold.

Sewage effluents, fertiliser run-off from fields and the organic wastes from intensive farming can all promote eutrophication by providing excessive quantities of plant nutrients. As a result, plants at the surface of the watercourse overgrow and shade the plants living deeper down. These then die and as they decompose they use up the oxygen in the water, rendering it virtually lifeless. Some of the surface plants may be blue-green algae which secrete powerful toxins.

The combination of warm, light summers and eutrophic conditions led to many algal blooms in reservoirs and lakes in Britain during 1989 and 1990. Some livestock died and several people were hospitalised as a result of contact with the poisoned water.

Eutrophication will only occur when all the required plant nutrients are present in excessive quantities. In some areas, for instance, there may be massive quantities of all but one substance and only if the deficient material is increased will the problem occur. The material in question is known as the limiting factor. In parts of the Norfolk Broads there are high concentrations of nitrates but phosphate seems to be the limiting factor, hence controlling phosphate inputs is the favoured option to prevent eutrophication. In parts of the Severn Estuary, where all nutrients are abundant, eutrophication does not occur because the water is too murky with sediments to prevent excessive plant growth: light is the limiting factor there.

FLUORIDES Fluorides are toxic compounds which are emitted during the smelting of aluminium and from some brickworks. They can accumulate in plants and pose a health hazard to animals which eat the plants. Although small amounts of fluoride are important for the development of healthy teeth, excessive amounts cause mottled teeth and damage to the bones, the joints of which tend to seize up, resulting in lameness. In addition to bone damage, fluoride poisoning has caused anorexia and anaemia in humans. Fluorides and other fluorine compounds have been used as rat poisons.

FORMALDEHYDE Formaldehyde is a gas emitted by various building materials such as urea-formaldehyde foam insulation and some brands of chipboard. It is also a component of photochemical air pollution. At fairly low levels it causes irritation to the throat, but some people are allergic to the substance and are affected at much lower levels. Formaldehyde is also considered carcinogenic, although authorities disagree about the magnitude of the risk.

FUNGICIDES Fungicides are chemicals used to control mildew, moulds and certain other plant diseases. They are also used as wood preservatives against wet and dry rot. Most are not of great environmental significance although some compounds have un-desirable short- and long-term effects: benomyl, for instance, reduces earthworm populations if used repeatedly. Mercury (q.v.) compounds are still used in some preparations although some of the most poisonous compounds in this group are no longer available. As mercury is a toxic heavy metal, the use of these materials is viewed with some concern.

Tributyl tin compounds such as TBTO, used as wood preservatives and anti-fouling compounds on boats, have caused damage in the marine environment – causing sex changes in shellfish – and are also thought to damage the immune system. They have been phased out in most of these applications in the UK but large ships are still permitted to use TBTO anti-fouling paints.

A group of fungicides which has given rise to concern in recent years is the dithiocarbamates (e.g. maneb, zineb and mancozeb). These are not especially dangerous themselves but break down to form the compound ethylene thiourea (ETU) which is carcinogenic. Worrying levels of ETU have been found in foodstuffs in the UK. Residue problems have also occurred with the unrelated fungicide tecnazene, which is used on potatoes. Some supermarkets now specify that the potatoes they buy should not be treated with tecnazene.

As with insecticides, fungicide use can lead to resistance in the organism against which it is used, and this is an increasing problem.

GENETIC ENGINEERING   The term genetic engineering covers a wide range of modern techniques for manipulating the features of organisms. Simple cross-breeding of crops and livestock could be regarded as a form of genetic engineering, but the term is usually used to mean modifying the genetic code – the system in the cell which determines how a living organism will grow, develop and behave.

Using these techniques it is possible to make bacteria produce human hormones, such as insulin, to introduce resistance to disease into plants and to create new organisms. The potential for food production, medicines and other chemicals manufacture, toxic waste treatment and many other applications is great. However, warning notes have been sounded about releasing 'synthetic' organisms into the environment as there is a risk that they may mutate into forms which could cause disease in humans, crops or livestock. If they are simultaneously resistant to conventional treatments the results could be disastrous.

In Britain the Health and Safety Executive currently monitors and controls all such experiments in this field and no problems have arisen yet. No wide-scale releases of genetically engineered species have yet taken place and safeguards must be built into the process before this happens.

GREENHOUSE EFFECT The surface temperature of the earth depends on a wide range of factors such as the amount of sunlight received, the amount of heat flowing up through the crust and the amount of heat reflected or radiated back into space. The first two factors are beyond human control but it is becoming increasingly apparent that the amount of heat trapped at the earth's surface is being increased by human activities.

The heat which is received from the sun is short-wave infra-red radiation and much of this is absorbed at the earth's surface. The surface then re-radiates some of the heat but this time it is in the form of longer wave infra-red radiation. Unlike the incoming rays, this radiation cannot pass easily through the atmosphere and is absorbed. As a result, the atmosphere warms up and heat is trapped near the planet's surface. This is known as the greenhouse effect since it was once thought – erroneously – that this is how greenhouses keep warm.

Not all the components of the atmosphere absorb the re-radiated heat. The principal gases doing so have been carbon dioxide, which is naturally present in the atmosphere at a level of approximately 300 parts per million, and water vapour. Levels were much higher in the geological past when the earth was warmer. Methane (q.v.) is another greenhouse gas and although this is also present naturally it is being augmented by human activities. Methane is much more effective at trapping heat than is carbon dioxide, while CFCs (q.v.) (which are completely synthetic) are some 7,000 times as effective as carbon dioxide. Nitrous oxide, also natural but added to by human activities, is increasing and has a long lifetime in the upper atmosphere, a property which it shares with CFCs. By the year 2030 it is expected that the effect of all these gases in terms of heat retention will be equivalent to a doubling of the carbon dioxide level when compared with the pre-industrial revolution figure.

The increases in these gases are relatively easy to measure providing that a site remote from local sources of contamination can be found. Less easy to measure is the resulting effect on temperature, since the earth's temperature varies considerably anyway and sorting out an overall increase from this variation is difficult. Nevertheless, a warming trend is becoming apparent in many areas of the world, although in some places there is evidence of cooling. A team of scientists at NASA, led by Jim Hansen, predicts that warming will be noticeable in most regions and by 2029 it will be warmer 'almost everywhere'. Estimates as to how much warmer vary, but one which has some credence is an average warming of 4 degrees C by 2030.

The predicted consequences of the greenhouse effect are

potentially catastrophic. The expansion of the oceans as they warm up and the melting of polar ice will raise sea levels so that many low-lying areas will be inundated. Much attention has been focused on the plight of cities but agricultural lands around the world – such as fertile deltas – and many islands will also be flooded. The climatic belts will shift towards the poles with many productive mid-latitude areas becoming deserts while tundras and northern plains become arable. The planet will be wetter, too, with more frequent and more violent storms. It has been suggested that the severe storms and warm winters experienced in Britain in recent years are evidence of global warming; although they could be co-incidental, if the pattern continues this explanation may prove to be true.

Ironically, although rainfall may increase in some places the warmer temperatures will make soils drier since water will evaporate more quickly, especially during growing seasons. Thus the water falling in the winter months will have little effect on soil moisture since by the time plants need it, it will be gone. This suggested effect is still the subject of much controversy – as are many of the predictions made about global warming – and it may be the case that crops will not need so much moisture if carbon dioxide levels are higher. This is because photosynthesis is more efficient under such conditions.

Exactly how – and indeed whether – the greenhouse effect will influence climate, humans and ecosystems is still a matter of prediction and complex computer modelling. Precise estimates cannot be made but most of the models agree that if heat-retaining gases continue to be released at predicted rates, then climatic change is inevitable, the only questions being when and by how much.

GROUNDWATER   Porous rocks under the ground harbour a vital source of clean water known as groundwater. It is stored in rock formations called aquifers and generally needs little treatment before use since it has been filtered by the rocks as it percolates downwards from the surface. Human activities, however, are threatening some of our groundwater supplies and as clean river water becomes scarce and demand for water increases this problem is likely to get worse.

The threats to groundwater come from materials used at the surface. Excessive quantities of nitrates, used as fertilisers or released by ploughing, percolate downwards. The same is true of nitrates from animal manures; for this reason the Dutch have imposed restrictions on intensive livestock farms to protect their water supplies. Some persistent herbicides, such as atrazine and simazine, are also reaching groundwater in unacceptable amounts while materials from landfill sites are imperilling a number of supplies. Solvents, carelessly disposed of on the ground or accidentally leaking from tips, have also polluted aquifers, although not on the scale seen in America's Silicon Valley where major solvent contamination has occurred.

US experience suggests that once a groundwater resource has become polluted it is virtually impossible to clean it up again. This means that it must either be abandoned or that the water extracted from it must be carefully treated before being used in public supplies. This can be expensive and is another example of the principle that preventing a problem is cheaper than solving it once it has arisen.

See also Drinking Water.

HALONS   Halons are gases and volatile liquids used in fire extinguishers and for other purposes where inertness and non-flammability are important. They are chemically similar to CFCs except that they contain bromine as well as, or instead of, chlorine. They are powerful ozone depleters and greenhouse gases and are to be phased out along with CFCs.

HAZARDOUS WASTE   There is no universally accepted definition of hazardous waste but a useful working one is 'waste that poses a risk to human health and the environment'. In practice most types of waste would fall into this category under one circumstance or another so the term is often restricted to industrial wastes which have particular properties. Hazardous wastes may be acutely or chronically toxic, they may cause birth defects or cancer, they may explode or catch fire or they may present a threat to water supplies or wildlife. Wastes normally discharged through chimneys or via pipes into rivers are not usually included in the definition.

Numerous instances of the mismanagement of hazardous wastes have occurred, from the dumping of cyanide drums near a children's play area in the English Midlands to the construction of community facilities on top of an old tip in Love Canal, USA, which resulted in toxic materials bubbling up through the ground and into basements. At the time of writing, drums of toxic and inflammable solvents are being excavated from an illegal dump on a farm near Bristol and the extent of this problem is yet to be determined. Once dumped, some materials can react together to produce new hazards – a tip near Edinburgh exploded for this very reason – and many materials will remain hazardous indefinitely.

Most of Britain's hazardous waste is dumped in landfills, some of which are known to be leaking. No site is completely leak-proof in the long term so complicated measures are taken at the newer, better managed sites to prevent the escape of noxious materials. Older tips were not so well managed and the locations of some are long forgotten, with the result that some now threaten drinking water supplies. A report from the waste management specialists at AEA Technology, Harwell, in 1990 showed that a third of the tips which received half the toxic waste landfilled in England were leaking and only half the sites in question had control measures installed to deal with the leakage.

Some hazardous wastes can be destroyed by chemical treatment – cyanides, for instance, are readily broken down by simple chemicals. Others can be incinerated, although there is a risk of air pollution. Several novel techniques are being developed, including new methods of incineration, electrical methods and sophisticated chemical treatments, but none is operating on a large scale in Britain at the moment. Increased charges for landfilling, as legislation and standards tighten up, will make some of these new options more financially attractive. In the meantime, a legacy of old badly-managed tips and a considerable area of contaminated land (q.v.) remains.

HCH   see Lindane.

HEAVY METALS   The heavy metals are a group of elements which have a number of chemical properties in common but whose biological roles range from the essential to the highly toxic. On the one hand, copper and zinc are essential trace elements required for the functioning of the body. On the other, mercury, cadmium and lead have no known biological role in the healthy organism and can cause damage at very low levels. Like any element, heavy metals cannot be destroyed and are cycled through the environment. They can be bioaccumulated by various organisms, particularly in aquatic systems, and a number of serious pollution incidents involving heavy metals have occurred. Pollution by heavy metals is discussed under their respective names.

HERBICIDES   Herbicides are chemicals designed to kill weeds and other plants. In terms of the way they work, they can be classified as contact or translocated. A contact herbicide – e.g. sulphuric acid – only kills the parts of the plants with which it comes into contact; whereas a translocated substance – e.g. glyphosate – is absorbed by the plant and carried throughout it, thereby killing the whole weed.

Most herbicides are not acutely toxic to humans or other animals, although there are some exceptions; paraquat is notoriously poisonous and has killed many people worldwide. Paraquat has also killed hares which have eaten contaminated stubble. The use of herbicides in parts of Scotland has drastically reduced the populations of insects upon which young partridges feed.

The chronic effects of herbicides have given rise to more concern since some have been found to be mutagenic or carcinogenic. The most widely publicised compound, in this context, is 2,4,5-T, a substance which was used widely in Vietnam for defoliation. It contains traces of dioxins (the material used by the US Air Force contained much more than preparations on sale for agricultural purposes) and there is evidence to suggest that it may cause cancer in its own right. Because of concern about the safety of 2,4,5-T its use for most purposes has been suspended in the US and restrictions exist in other countries. The British government still believes that it is safe and has refused to ban it, although manufacturers have voluntarily removed 2,4,5-T from home garden products. 2,4,5-T is a member of the phenoxyacetic acid groups of herbicides and Swedish studies have linked the use of this group as a whole with a specific type of cancer in forestry workers.

Most herbicides are not particularly persistent in the environment although one group, the triazines, does leave residues for some considerable time. Drinking water (q.v.) supplies in parts of England are contaminated by traces of atrazine, a member of this group. It is used extensively by arable farmers, local authorities and British Rail to kill all types of weeds and is washed down through the soil to the water table.

The widespread use of herbicides, coupled with other intensive agricultural methods, has led to the disappearance of many native wild flowers from areas of Britain. Some of these are regarded as weeds – e.g. poppies – and are sprayed directly when they occur in fields but others are not a threat to productivity, living in verges and hedgerows which are sprayed incidentally. A traditional meadow contains dozens of different plants each of which provides a different balance of nutrients for grazing animals. Modern grasslands are often heavily fertilised, treated with herbicides and may contain only one strain of one species of ryegrass, a considerably impoverished system when compared with the meadow alternative.

HYDROCARBONS   Hydrocarbons are chemicals composed solely of hydrogen and carbon. Crude oil consists mainly of

hydrocarbons and the group includes thousands of different substances ranging in properties from gases such as methane and propane to solids such as waxes and bitumen. Apart from the consequences of oil spills into waterways and the sea, the main environmental concerns about hydrocarbons relate to the lighter, more volatile compounds. Methane (q.v.) is a greenhouse gas and the volatile hydrocarbons used as fuels (petrol) and solvents can interact with sunlight and other materials in the air to promote the formation of ozone (q.v.) and produce photochemical smog. Some hydrocarbons (e.g. benzene and many polycyclic aromatic hydrocarbons) are carcinogenic.

HYDROGEN CHLORIDE   Whenever plastics containing chlorine (e.g. PVC) are burned the gas hydrogen chloride is produced. This is an irritant choking substance and dissolves in water to form hydrochloric acid. It will attack the lining of chimneys and contribute to acid rain, so incinerators and other plants emitting hydrogen chloride should be fitted with scrubbers to remove the gas. The gas is also harmful to plants and its emission from industry in the north-west, and subsequent damage to crops and livestock, led to the first major industrial pollution control legislation in Britain, the Alkali Acts.

INCINERATION   Incineration is the destruction of wastes by burning and is one of the oldest methods of waste disposal. About 10% of Britain's municipal waste is incinerated but this technique is more popular in other countries; in Switzerland and Denmark some 70% is dealt with in this way. A small proportion of hazardous waste and much of the clinical waste produced in Britain is also burned. In some instances useful energy can be reclaimed from incinerated wastes. The Edmonton incinerator in north London produces a useful quantity of electricity for the national grid while other schemes provide hot water for district heating.

The main environmental problem with incineration is the air pollution which may be produced. Ironically, municipal waste incinerators can be more polluting than specialist hazardous waste plants which include sophisticated pollution control measures. Clinical waste incinerators on hospital premises have been extremely polluting in the past since they were covered by Crown immunity and not subject to normal air pollution controls. More clinical waste is now going to specialist commercial plants and Crown immunity has been lifted.

Two main air pollution problems arise from incinerators: metals, and incompletely burned organic compounds. Gases such as sulphur dioxide and nitrogen oxides may also be problematical in some circumstances and materials containing chlorine compounds produce the acidic gas hydrogen chloride on combustion. Burning wastes containing cadmium, lead and mercury will cause the release of these metals, usually as particles, from the chimney stack and contamination of the surrounding land may result if measures are not taken to extract the pollutants from the effluent gases.

Incomplete combustion results in smoke and, perhaps more importantly, the generation of dioxins and dibenzofurans (q.v.). These substances are released when temperatures in the furnace are not high enough to destroy them and when there is insufficient oxygen available for complete combustion. Poor mixing of the burning wastes and too short a time spent in the burning zone can also contribute to the problem. Many domestic waste incinerators do not achieve the necessary combustion conditions and dioxins are released as a result.

The incineration of hazardous wastes has been controversial and allegations that the process has caused disease and birth defects in livestock and humans have been made. Of particular concern is the incineration of PCBs and other chlorine-containing wastes which are liable to form dibenzofurans and dioxins as described above. While standards at some plants were undoubtedly lower in the past, evidence of serious health effects is inconclusive. If the sophisticated control systems and pollution prevention measures installed on modern incinerators work properly there should be no problems with the incineration of hazardous or domestic wastes, but many residents living around such plants remain unconvinced.

Some wastes have been burned at sea by specialist incineration ships but this practice is being phased out, partly as a result of campaigning by Greenpeace. Elevated levels of heavy metals have been found in sediments in the North Sea below marine incineration sites and there have been instances of clouds of hydrogen chloride, which is not scrubbed out of the waste gases, being blown towards the shore. Process control on the ships is difficult and concern has been expressed about the formation of dioxins and dibenzofurans when chlorinated wastes are burned.

INDOOR AIR POLLUTION Pollution is generally thought of as an outdoor phenomenon but it affects us indoors as well. Apart from exterior air pollutants entering buildings, many substances produced from building materials and domestic activities can be harmful. Offices and other commercial premises also generate their share of indoor pollutants and the term 'sick building syndrome' has been coined to describe buildings wherein a combination of factors makes working there unhealthy.

A major pollutant in many buildings is tobacco smoke. This adds particulates (q.v.) to the air and also some carbon monoxide, although the latter is not particularly significant in most cases. The particulates contain carcinogenic materials and present a risk to the health of non-smokers as well as smokers. Young children from smoking families suffer a greater frequency of lung problems in their early years than do the children of non-smokers.

Other combustion processes also release harmful substances and this can be dangerous if ventilation is inadequate. Any fuel-burning heating appliance which is not vented to the outside – e.g. paraffin stoves and portable gas heaters – can cause a build-up of combustion gases as well as condensation of water vapour on cold walls and ceilings. Gas cookers can also emit these gases and adverse health effects have been associated with their use in poorly ventilated circumstances.

Building materials can emit gases and vapours in normal use. Other substances can pollute the air in homes under exceptional circumstances. Urea-formaldehyde foam, used for insulating cavity walls, releases formaldehyde as it cures; chipboard often gives off this gas as well since formaldehyde-based adhesives are used in its manufacture. Formaldehyde is irritant, an allergen and is believed to be carcinogenic. PVC furnishings and floor tiles may release quantities of vinyl chloride (VCM)(q.v.), while a range of volatile compounds may be emitted by carpets. In most cases emissions of these pollutants decline with time as the insulating foams set and uncombined volatile materials in other products disperse. This means that good ventilation is needed initially but becomes less important over time. This is not the case with radon (q.v.) which is emitted constantly from some traditional building materials such as certain granites and also diffuses up from the ground in some areas.

Disturbance to building materials can cause indoor pollution problems, the most serious being the dust produced when leaded paint is sanded down dry, especially with power sanders. Blowlamps are dangerous too since they vaporise some of the lead. Older houses often contain large amounts of leaded paint under layers of more modern lower-lead products. Careless removal of this material can result in lead poisoning and the dust produced can contaminate soft furnishings and surfaces throughout the house. Asbestos is also dangerous if disturbed; drilling wall panels or sanding down textured paints containing asbestos releases the carcinogenic fibres into the air.

Other DIY activities are not without risk. Some paints, especially gloss finishes, release potentially harmful solvents as they dry and white spirit fumes can also be hazardous. If these fumes contact a flame, such as a gas ring, they are partly combusted to produce a range of irritant materials. Solvents from other products, including adhesives and even some wax polishes, can also be harmful if breathed to excess.

In office buildings a range of factors combine to produce sick building syndrome. Some of these are not fully understood. A major factor seems to be the air conditioning systems which tend to be enclosed, recycling the same air several times. The humidity is controlled by the system and sometimes the air may become contaminated by micro-organisms growing in water as a result. Materials released from furnishings, solvents used in printing and ozone from photocopiers are also potential sources of discomfort, as is tobacco smoke in those offices which still permit smoking. Lighting and psychological factors are also thought to play a part and the overall picture is very complex.

Ironically, one of the factors making indoor air pollution worse is energy conservation. Many houses are now much better draught-proofed than they used to be and this means that the air is changed less frequently. As a result, indoor pollutants can build up to higher levels than they would in draughty premises. A balance must therefore be struck between ventilation and conservation, although before this can be achieved it makes sense to minimise sources of indoor pollution.

INORGANIC COMPOUNDS
Inorganic compounds are those consisting of elements other than carbon. Some simple carbon compounds, such as carbon monoxide, carbon dioxide and carbonates are considered to be inorganic. The term 'inorganic' is also applied to farming methods which involve the use of inorganic fertilisers such as nitrates and phosphates made synthetically.

INSECTICIDES
Insecticides are chemicals used to kill insects and similar pests. The majority are used in agriculture but large quantities are used for woodworm treatments and in domestic gardens. In the past, some of the chemicals used have been highly toxic and have wrought havoc on wildlife. Some of the worst compounds are now banned in Britain but there are still problems with permitted products and the use of some banned compounds continues illicitly.

There are many different types of insecticide but a few main categories can be identified. The most notorious group is the organochlorines (chlorinated hydrocarbons) and this includes DDT, dieldrin, lindane and chlordane. Most compounds in this group are toxic to highly toxic and they are persistent in the environment. Organochlorines have caused human fatalities and untold damage to wildlife around the world. In the developed world most organochlorine insecticides are banned or tightly restricted, although lindane and HCH are used freely in some places including Britain. In the developing world, organochlorines are still used on a wide scale for agricultural and public health purposes, partly because they are comparatively cheap.

A second major insecticide group is the organophosphorus compounds. These include some extremely poisonous materials, such as parathion and mevinphos, although others are less virulent. Organophosphorus insecticides do not normally persist in the environment, the main problem with their use being the deaths of non-target species. They have also had serious effects on humans, causing many fatalities and instances of non-fatal poisoning. These compounds affect the nervous system by inhibiting an enzyme called acetylcholinesterase and some people are very sensitive to this type of compound. Such people should avoid contact with organophosphorus insecticides.

Another group of insecticides is the carbamates. These resemble the organophosphorus group in their action – they inhibit acetylcholinesterase – but are different chemically. They are not persistent in the environment but some, such as aldicarb, are highly toxic.

Most other compounds are not easily classified into groups since they vary so much in their chemical structures. There is a range of chemicals called the pyrethrins which are either derived from chrysanthemum-like flowers or made synthetically and a number of other low-persistence compounds such as nicotine and derris. These last two are naturally occurring and their advantage from the environmental point of view is their low persistence. However they are by no means harmless, since derris is highly toxic to fish and nicotine is very poisonous to most animals.

The problems of insecticide use may be short-term or long-term. If a non-selective insectide is sprayed indiscriminately it not only kills the target pest but also the predators of that pest such as spiders and ladybirds. Repeated spraying so depletes the predator populations that the pest can multiply even faster and yet more insecticide is needed. In this way, pests can be created; removal of predators means that creatures which were formerly present at very low levels can multiply and become a serious nuisance. Non-target species can be beneficial in other ways – orchards depend on bees for fruit pollination but these insects are often killed by sprays.

Repeated spraying of the same type of insecticide leads to another problem: pest resistance. The target pests evolve an immunity to the chemical so that ever-larger doses are required and more environmental damage is done. Eventually the insecticide becomes useless and another type of compound must be tried. A pest which is resistant to one type of organophosphorus compound is often resistant to others in the group, so finding an alternative is not easy.

Persistent insecticides pose other risks: bioaccumulation and residues on food. The organochlorine compounds have caused much damage to populations of birds of prey, including owls, otters and other species at the top of food chains as their residues are passed from prey to predator. Such is the persistence of DDT that it is difficult to find anywhere that is uncontaminated, although in most cases this contamination is not of significance to health. Even short-lived compounds can harm predators if the prey absorbs a large dose just before being eaten – a process called secondary poisoning.

The more persistent materials can also leave residues on food, despite the requirements for an interval to be left between spraying and harvesting. Some imported foods have been found to be contaminated by materials banned in Europe and it is ironic that Western countries have exported chemicals banned on health grounds at home and then imported food from the Third World contaminated by the same materials. This is sometimes referred to as the 'circle of poison'. The types and dosages of pesticides which may be sprayed on foods are regulated, but errors and deliberate breaches of the regulations sometimes occur. In July 1985 aldicarb was accidentally used on Californian watermelons and people who ate them suffered nausea, diarrhoea and trembling as a result. In Britain, excessive levels of dieldrin have been found in cows' milk on occasions, and the organophosphorus compound dimethoate tends to exceed the maximum recommended limit in some types of lettuce.

Most modern agricultural systems are dependent on the use of insecticides and other pesticides and the sudden removal of these compounds would lead to a massive upsurge in crop damage. Organic farmers, however, achieve respectable yields without their use. Research is currently directed, in the main, towards finding better insecticides which do not harm non-target species, and some success has been achieved in this respect. In the long term, integrated pest control (see below) is likely to become more widespread with pesticides playing a comparatively minor role. Indeed enthusiastic organic farmers believe that we can do without these compounds entirely.

INTEGRATED PEST CONTROL   Integrated pest control is a modern, sophisticated approach to keeping damage from pests down to a minimum. It is impossible to eliminate pests completely. An integrated pest control programme uses a wide range of techniques and requires thorough knowledge of the pest, the crop and the responses of both to the environment.

The first line of approach is to consider whether the crop can be made less susceptible to attack from the pest. Selective breeding for protective characteristics, such as tougher leaves, could be an initial step. Planting crops in different ways or at different times may be useful; traditional crop rotations help to keep down soil-living pests, for instance, and growing plants which repel insects between the rows of crops may also reduce losses.

Biological control (q.v.) techniques, often a part of integrated pest control, can involve releasing parasites or predators of the pest or trapping it with lures that resemble natural pheromones (chemicals which attract members of the opposite sex). Pesticides may have a place in integrated pest control but the quantities used are generally small and the methods used to apply them are precise, with measures taken to minimise damage to beneficial species.

Integrated pest control is not easy, requiring thorough research and monitoring before and during its use. It can be much more effective than conventional mass spraying methods and is often much cheaper. Furthermore it greatly minimises the risks to the environment presented by pesticides and is unlikely to lead to pest resistance. Most of the successful integrated pest control programmes have been carried out overseas, although British organic farmers practice these techniques – without the pesticides.

INVERSION CONDITIONS   see Temperature inversions.

LANDFILLS   The majority of Britain's domestic and industrial waste is disposed of in landfills (tips). This practice has, in the past, resulted in air and water pollution, contamination of land and serious nuisance to people living nearby. Fires, explosions and the release of toxic gases and dusts have occurred and pests such as flies, rats and gulls are an ever-present problem on domestic waste sites.

Some of these problems can be overcome by good management.

Vermin and nuisance can be controlled by covering wastes with soil after each day's tipping and careful control of the wastes accepted should prevent unwanted chemical reactions leading to pollution or explosions. Methane can be collected and used as a source of energy rather than allowed to escape or accumulate in buildings. A suitable choice of site, which may need to be lined and the leachate produced collected for treatment, can reduce the risk of water pollution while the pre-treatment of industrial wastes can reduce the migration of toxic materials.

Some fundamental problems are less easy to deal with. First, there is a legacy of old sites which were operated to very lax standards and were often in quite inappropriate locations. Some of these are known to be threatening groundwater supplies while the locations of others are not even identified.

Second, landfilling some types of toxic waste does not remove the problem, it simply stores it up for future generations. No landfill is completely secure against leakage in perpetuity and sooner or later some of the materials dumped will escape. If they are degradable in the environment there may not be a problem – indeed processes in the tip may break them down before any escape is likely – but persistent materials such as heavy metals can be reconcentrated by living organisms, possibly to dangerous levels.

A third point is that landfilling wastes often represents a waste of resources. Valuable metals, glass, paper and plastics are all recyclable, in theory at least, and even if they cannot be reused as materials some waste components can be used as a source of energy in refuse-derived fuel plants. Recycling materials usually involves the consumption of less energy than the production of new materials and this has implications for pollution generation and the greenhouse effect.

In many countries, landfill standards are being improved. The US has banned the landfilling of most hazardous wastes and several EC Directives will lead to restrictions on the types of wastes dumped and the manner in which tips can be operated. As costs rise, other more environmentally-friendly methods of waste disposal will become more economic, but landfill looks like remaining a major disposal route for domestic and industrial wastes for many years to come.

LEACHATE   Leachate is the liquid which seeps from wastes deposited in landfills. It contains a variety of polluting materials including ammonia, toxic metals, chlorinated hydrocarbon solvents and other toxic materials deposited in the waste. Leachate will usually have a very high BOD (q.v.) and can damage aquatic ecosystems and pollute groundwater. The composition changes as the tip ages, depending on the chemical and biochemical reactions which have occurred. Leachate can be treated although it is not current practice to do so at most British sites despite some notable successes. Modern landfills in Germany and the US do include leachate treatment facilities, however, and impending EC legislation may make it the norm throughout Europe. Leachate generation can be controlled by restricting the amount of rainwater entering the tip and by lining the site. Many older landfills in Britain, containing domestic and industrial wastes, are leaking and their leachate may threaten groundwater resources: dealing with such sites will be extremely expensive – if, indeed, they can be dealt with at all.

LEAD   Lead is a poisonous heavy metal which has been used by humans for thousands of years. Batteries, solder, roofing materials and petrol still contain lead as do some specialist paints. Many thousands of buildings have old lead pipes and layers of paint containing the element, sometimes at alarmingly high concentrations. Modern paints are generally unleaded, as is some petrol, but a legacy of long-term lead use contaminates our soil, drinking water and the air.

Acute lead poisoning has been known for centuries but is fortunately uncommon these days. Every year a few children are admitted to hospital with lead poisoning as a result of chewing old paint, but fatalities are rare. Of more general concern is the fact that lead levels in the population as a whole are raised far above the natural level and there is evidence that some children are intellectually impaired as a result. Levels of lead which do not produce any physical signs or symptoms of poisoning have been found to impair the functioning of the brain and the performance of children in a wide range of psychological and educational tests.

Levels of lead in the general environment are beginning to decline as a result of the introduction of unleaded petrol. Nevertheless they are still much higher than they should be and while leaded petrol is still in use it will contaminate people who breathe it and also enter the food chain as it is deposited from the air onto growing crops and food preparation surfaces. In addition, lead pipes add the metal to drinking and cooking water. Old lead paint is harmless if left undisturbed – and unchewed – but the use of blowlamps, high temperature hot air strippers and abrasives to remove it can lead to massive contamination of the home by lead dust. Even if the use of lead was stopped tomorrow, much human-produced lead would remain in the environment for centuries.

LEUKAEMIA CLUSTERS   Leukaemia is a form of cancer which affects the blood. It takes several forms but certain types are known to be caused by exposure to ionising radiation. It is a comparatively rare disease but in some parts of Britain there have been unusually high numbers of cases detected in children and young people. These are known loosely as leukaemia clusters and their especial significance is that some of these clusters occur around nuclear establishments. The first such cluster was found around Sellafield and another has been detected near the experimental nuclear site at Dounreay in Scotland. However not all nuclear sites exhibit clusters of leukaemia and not all clusters are associated with nuclear sites.

There are many problems involved in interpreting these data. First, a rare disease like childhood leukaemia may not be distributed smoothly across the country. Clusters could occur by chance, although the probability of the number of cases occurring around Sellafield being purely a chance phenomenon is some 3 million to one against. Second, cases may be missed by choosing the wrong period of time in which to record their numbers, or by ignoring population movements. Someone moving away from the area may have contracted the disease locally but would appear as a statistic somewhere else: similarly someone may enter the area already suffering from the disease and add to the statistics locally.

The arguments against the theory that the nuclear plants are responsible for the increases in leukaemia cases are based on beliefs about how much exposure to radiation will cause a given number of leukaemia cases. Nuclear industry sources say that the amounts of material discharged from the plants are not sufficient to cause that many cases of leukaemia, therefore the plants are not responsible. This prompts the questions as to how reliable is our understanding of the mechanisms involved and how reliable are the data on emissions. It is known that misinformation on discharges was supplied to the Black committee which looked at leukaemia around Sellafield and it is also known that much higher levels of radioactivity were found on beaches near Dounreay than were predicted. Our understanding of how cancer is induced is far from perfect and we do not know whether unborn children or babies are particularly sensitive to the exposures encountered.

No satisfactory alternative explanation for the leukaemia clusters around nuclear plants has been advanced, although there is growing evidence that viruses may be involved in some types of leukaemia. Increases in other types of cancer around other nuclear plants, and in workers with nuclear materials, have added weight to the theory that the plants are responsible.

LINDANE Lindane is an organochlorine insecticide used widely in agriculture, in the garden and as a woodworm treatment. It is also used to kill wasps and carpet moths. It is a particularly pure form of one of the materials which makes up the material HCH. Both HCH and lindane have been associated with many cases of poisoning, in humans and wildlife, around the world. The substance has also been linked with leukaemia and anaemia and for this reason pressure has grown for its withdrawal from the market. It is restricted or banned in Scandinavia, the US, the USSR and New Zealand, but is still widely available in the UK. In 1989 the UK government announced that the safety of lindane would be reviewed but it has remained on sale in the meantime.

MARINE DUMPING The dumping of wastes at sea has been practised for decades and is now in decline. By the end of 1993 no more industrial wastes will be dumped at sea by Britain and the marine disposal of sewage sludge is scheduled to finish by 1999. The problems posed by marine dumping are twofold: (1) localised transient problems may occur, as when too much material capable of causing eutrophication (q.v.) is dumped under conditions of poor dispersal; (2) persistent materials may subsequently bioaccumulate when introduced into the seas.

There is evidence that localised problems do occur in some areas where sewage sludge is disposed of and Greenpeace has marshalled an impressive case documenting the harm done to fish by the dumping of acid wastes from titanium dioxide (q.v.) production. Persistent materials such as heavy metals do seem to accumulate in some areas. Liverpool Bay, where much waste has been dumped and which also receives the polluted waters of the Mersey, supports fish which contain high levels of metals and PCBs.

A further problem is the legacy of chemical warfare agents dumped at sea after the two world wars. Chemicals such as mustard gas and phosgene are involved and no-one knows how much material was dumped or where it is.

The marine disposal of wastes in the North Sea and North Atlantic area is regulated by the London Dumping Convention. Under this convention some materials are banned from being dumped while others may be disposed of at sea only under licence. Britain acceded to international pressure to phase out the dumping of most wastes but wanted to retain the right to dump some nuclear wastes, a practice which has been suspended as a result of actions by environmental groups and trades unions. At the 1990 meeting of the London Dumping Convention all undersea nuclear waste disposal was banned, which leaves several countries with the problem of what to do with old nuclear submarines.

MERCURY Mercury is a volatile and extremely toxic heavy metal. It is liquid at room temperatures, hence its use in thermometers, and forms a range of compounds all of which are toxic.

Mercury poisoning has been known for centuries. The expression 'mad as a hatter' derives from the mental illness suffered by hatters who used mercury in the preparation of felt. Acute mercury poisoning is characterised by pallor, stomach pains, delirium, coma and death. Low-level chronic mercury poisoning is less easy to detect. The victim may suffer from confusion, flushing, shyness, headaches, fatigue and tremors: staff at a dental surgery exhibited some of these symptoms when a pool of mercury built up under an oven and the windows were shut for the winter.

The toxicity of the metal depends on its form and how it is absorbed. Mercury metal can be swallowed and little will be absorbed, but if it is inhaled it can be lethal. Inorganic (q.v.) mercury compounds are toxic but some are not very easily absorbed. Organic (q.v.) compounds, such as methylmercury, are very easily absorbed and enter the brain readily, causing serious damage.

The most infamous environmental disaster involving mercury occurred at Minamata Bay, Japan, in the 1950s. Wastes containing mercury, some of it methylmercury, were discharged into Minamata Bay, the source of the fish forming the basis of the local diet. The mercury bioaccumulated and high levels ended up in the fish and hence in humans. According to official figures 46 people died and 68 were permanently disabled by organic mercury poisoning; some of the disabilities resulted from the exposure of babies before birth. About 700 people were poisoned in all.

Although some of the mercury discharged into Minamata Bay was the more toxic organic form, which bioaccumulates particularly easily because it is soluble in fat, some of the methylmercury was made in sediments by the action of bacteria and organic matter on inorganic mercury. Thus natural processes can make a pollution problem worse.

Mercury in effluents is now tightly controlled and discharges to the sea around Britain are falling. Sewage sludge from London, which used to contain significant quantities, now contains less since the water authority clamped down on the discharges of the metal into public sewers.

Nevertheless, mercury still reaches the environment from a variety of sources. Burning coal and refuse releases the substance into the atmosphere and conventional pollution control techniques are not effective at removing it from flue gases because it is so volatile. Long-life batteries contain mercury which will be released when the batteries are incinerated or dumped, although some manufacturers have cut levels substantially in recent years. Hearing-aid batteries are often mercury-based and could be recycled, although few are. There are alternatives to mercury thermometers, although it is less easy to replace in other scientific instruments. Mercury has also been used in children's toys (the Mercury Maze), a practice which has been condemned as a health hazard, and some toy batteries contain the metal. Some fungicides still contain mercury and Friends of the Earth is calling for these to be banned.

METHANE   Methane is the simplest hydrocarbon (q.v.) and is the principal component of natural gas. As well as occurring in oil and gas deposits it is produced by biological processes in pond muds, paddy fields, refuse tips and the guts of many living organisms from termites to cows.

Methane causes two environmental problems: one local and one global. On a local basis, methane from rubbish tips can pose a fire and explosion risk. It seeps through fissures in the ground and can accumulate in the basements of buildings. A house at Loscoe, in Derbyshire, exploded in 1986 when a build-up of methane ignited. Properly managed tips now include measures for ventilating and flaring off the methane while some sites use it as a fuel, but some 1,300 British tips are thought to present a risk of explosion. Methane has long been a problem in coal mines; its ignition has been responsible for countless pit disasters. It was announced in 1990 that the Derbyshire village of Arkwright is to be moved in its entirety because methane from old coal mines threatens the houses constructed above.

The global problem is the greenhouse effect (q.v.). Methane is 60 times as effective as carbon dioxide at trapping the sun's heat, and levels in the atmosphere are rising – by 1% per year during the 1980s. Human activities are assisting in this increase in several ways. Creating more rice paddies to feed growing populations adds to methane emissions while clearing tropical forests provides an ideal environment for termites to multiply. The world's population of cattle is increasing, as is the amount of domestic waste capable of producing the gas. Even tarmac, the basis of most road surfaces, releases methane when heated by the sun and British Gas, which has run a series of expensive newspaper advertisements about its environmental responsibility, has no precise figure for how much methane leaks from its distribution mains.

MOTOR VEHICLES Motor vehicles have replaced the burning of coal as the principal source of air pollution in most cities and towns. Most of the pollution emerges from the exhaust but petrol fumes also leak from the fuel tank and particles of rubber, which may be carcinogenic, are rubbed off tyres. Some brake linings contain asbestos and this adds to the background levels of this material.

If fuel – petrol or diesel – burned completely there would be no emissions of carbon monoxide or hydrocarbons from the exhaust although nitrogen oxides, formed when oxygen and nitrogen in the air are heated to high temperatures, would still be produced. Fuels never burn completely, however, so all three pollutants are released. Some 85% of the carbon monoxide emitted in Britain comes from motor vehicles as does 45% of the nitrogen oxides and 28% of the hydrocarbons. The fact that these emissions are often localised increases their polluting potential.

Of course, any carbon-containing fuel produces carbon dioxide when burned and motor vehicles account for 16% of the carbon dioxide produced in Britain by human activities other than respiration. Any sulphur in the fuel is converted to sulphur dioxide and legal limits exist to control the amount of sulphur in diesel.

The materials emitted by motor vehicles may be hazardous in their own right but in combination they can be much worse. Nitrogen dioxide, ozone, PANs and a variety of other secondary pollutants are produced in photochemical smogs (q.v.) if the sun is bright enough; these can have serious effects on health. Diesel smoke, which is dark and soiling, contains polycyclic aromatic hydrocarbons, some of which are carcinogenic. Laboratory studies suggest that they may become more so when reacted with ozone.

One particularly hazardous pollutant from motor vehicles is lead (q.v.), emitted from cars and motorbikes running on leaded petrol (lead is not added to diesel or liquefied petroleum gas). When lead levels in petrol were higher in the 1970s, and all petrol was leaded, some 90% of the lead in the urban atmosphere came from motor vehicles and the roads network ensured that atmospheric lead fallout was well distributed over soils and growing crops. The introduction of unleaded fuel and the reduction of the amount of the metal permitted in leaded petrol is reducing the amount emitted to the air by cars but, many would argue, not fast enough since 80% of atmospheric lead still comes from vehicles.

There are some technical fixes for motor vehicle pollution, unleaded petrol being the most obvious one. Catalytic converters (q.v.) can greatly reduce the emission of hydrocarbons, carbon monoxide, and nitrogen oxides. Better design and maintenance of engines can also help to control emissions. The only way to deal with carbon dioxide emissions, however, is to design cars to be more fuel efficient and use them less often. Many environmentalists now see restraint on private transport as being the only way of reducing carbon dioxide emissions from this sector, but the idea has not, so far, found favour with government.

NITRATES

Nitrates are simple chemical compounds which are found naturally as part of the nitrogen cycle. They are essential for plant growth and the nitrogen in them is incorporated into proteins, hence providing all the protein, directly or indirectly, used by animals. Nitrates form when bacteria act on ammonia (q.v.) in the presence of oxygen and are also produced from gases which form during lightning storms.

Pollution problems with nitrates occur when humans upset the balance within the nitrogen cycle. Too much nitrate in the wrong place can cause an excessive uptake by some plants, air pollution, eutrophication (q.v.) and the contamination of drinking water. The excessive uptake of nitrates by plants means that they cannot use the substance to make protein: hence some is stored. This, in turn, is consumed by humans or animals and adds to the nitrate burden in the diet – which may be hazardous as is explained below. In some instances – for example spinach – the nitrate is converted into the more toxic nitrite on storage.

Air pollution results from excessive nitrate use because of bacterial action. Nitrate is converted into the gas nitrous oxide which moves from the soil into the atmosphere. There it may be oxidised into nitrogen dioxide and contribute to acid rain. If it is not oxidised it may still cause problems, since it is a greenhouse gas.

People add nitrates to the water cycle in two main ways: in sewage effluents and as fertilisers. Sewage effluents are responsible, on average, for about 50% of the nitrates in British rivers but the exact proportion will vary greatly from place to place. Sewage treatment processes convert ammonia to nitrate and it is rare for nitrate removal equipment to be installed.

Nitrate fertilisers and, to some extent, animal manures contribute the remaining 50% on average but in some areas, such as East Anglia, the proportion is higher. Nitrates from fertilisers and manures are also threatening groundwater since much of the fertiliser applied is not taken up by plants and seeps into the ground. Other agricultural practices, such as ploughing up permanent grassland, also add nitrates to the groundwater and there is often no direct correlation between the amounts applied – if any – and the quantities moving down towards aquifers. One possibility is that the addition of nitrate fertilisers stimulates the soil micro-organisms to mobilise reserves of nitrogen already held in the soil: thus the polluting effect may be indirect. Nevertheless, there are some areas where excessive fertiliser use is likely to render groundwater unusable for public supply, although the extent of the problem may not be apparent for many years: it takes decades for the substance to move down to the water table.

The potential hazards from nitrates in drinking water are twofold. High levels taken in by babies can cause a condition called methaemoglobinaemia where nitrite, produced from nitrate in the gut, combines with haemoglobin and prevents it from carrying oxygen. This is similar to carbon monoxide poisoning (q.v.). Toddlers and older people are not at risk as their haemoglobin is different from that in young babies and the bacteria in the mouth which help to convert nitrate to nitrite become less abundant as the baby grows.

The second potential problem is that of stomach cancer. Nitrite can combine with substances in the diet (amines) to form compounds called N-nitrosamines. Some of these are carcinogenic and there is concern in some quarters that excessive levels of nitrate in the diet could lead to stomach cancer. So far, much of this risk is theoretical as studies of populations exposed to different levels of nitrates have not demonstrated a link between the contaminant and stomach cancer. Nevertheless, many would argue that it is prudent to limit nitrate intake in case the risk becomes reality.

Nitrate levels in drinking water have been controlled by some water companies in the past by blending water from different sources to produce an acceptable dilution. Unfortunately, suitably uncontaminated sources are becoming unavailable in some areas and other methods have to be used. Nitrates can be removed by treating the water but this is expensive and the costs fall on the water consumer rather than the polluter, so the idea of a fertiliser tax has been suggested. This has not been adopted by the government – but the idea of nitrate exclusion zones has. Within such zones, the application of nitrate fertilisers will be restricted and compensation may be paid to farmers. Britain has taken action following a prosecution in the European Court of Justice initiated by Friends of the Earth. Previous plans would have resulted in the UK failing to comply with an EC Directive on drinking water quality.

In 1989 some 1.6 million people in Britain received water with more than the EC limit of nitrate, the worst affected populations being in farming areas. Nitrate levels are still rising.

NITROGEN OXIDES   The gas nitrogen, which forms four-fifths of the atmosphere, does not easily combine with other materials, but when it does the chemicals formed are often of biological and environmental interest. Among the simplest of these compounds are the nitrogen oxides where nitrogen combines with either one, two or half an oxygen atom per atom of nitrogen.

These gases form naturally in the environment as a result of biochemical and geological processes. Problems occur when human activities increase the concentrations, locally or globally, so that equilibria are disrupted.

On a global basis, nitrous oxide – known as laughing gas – is one of the gases responsible for the greenhouse effect. Artificial fertilisers are adding to the quantities produced naturally, so that the concentration of nitrous oxide in the air is increasing at around 0.2% per year.

More locally, nitrogen oxides from vehicle and industrial emissions cause pollution problems. Vehicle engines emit nitrogen oxides mainly in the form of nitric oxide. This is converted in the atmosphere into nitrogen dioxide, a toxic and irritant gas. Nitrogen dioxide affects the lungs, causing irritation, bronchitis and pneumonia. Resistance to infection is also lowered. In general, average values for nitrogen dioxide in the air in Britain are below those at which health effects would be expected. However in central London some very high peaks have been measured, above the point at which health effects may occur, and levels appear to be rising. World Health Organisation guidelines are exceeded in a number of locations in central and southern England. On health grounds, nitrogen oxides are likely to be more significant as sources of ozone (q.v.) and photochemical smog (q.v.) rather than as poisons in their own right.

In addition to its specific effects on health, nitrogen dioxide is involved in the production of photochemical smog in sunny urban areas since it can assist in the formation of ozone. In the stratosphere, nitrogen oxides are thought to be capable of depleting ozone and concern was expressed about the environmental effects of large fleets of supersonic aircraft for this reason. Supersonic flight is still only a luxurious curiosity at the moment so this threat has not been realised. Ironically, nitrogen dioxide may, in small quantities, exert a protective effect on the ozone layer by combining with some of the harmful chlorine from CFCs.

When nitrogen dioxide dissolves in water it forms an acidic solution. In Europe about a third of the acidity in rainfall comes from nitrogen oxides. Motor vehicles are the main source but power stations and other large combustion plants are also significant.

NOISE   Although noise does not threaten the safety of the planet, it is a growing urban pollutant as well as a major source of industrial ill health. Noise levels in towns are rising and more and more people are exposed to excessive levels. This is mainly as a result of more traffic, the widespread and inconsiderate use of powerful musical equipment such as portable tape players, increased air traffic and, in some instances, the enthusiasm for do-it-yourself. In factories, regulations and codes of practice for noise control to protect employees exist, but are not enforced adequately. Prosecutions for breaches are extremely rare.

Prolonged high levels of noise cause deafness and aggravate the hearing loss which occurs naturally with age. This is of significance in industry but the levels concerned do not usually occur in the general environment for any length of time. Lower levels cause loss of sleep, irritation, poor concentration and seriously reduce the quality of life. Any health effects are normally psychological rather than physical.

While it is possible to control noise emitted by, for instance, cars or lawnmowers by means of standards it is less easy to control overall exposure to noise since the use of cars is increasing. Environmental health officers do have powers to abate noise from parties and other noisy activities but, in practice, this is difficult to achieve satisfactorily: apart from any technical problems, there are simply not enough environmental health officers.

NORTH SEA   The North Sea has been described as one of the world's most polluted seas. It is certainly one of the world's busiest in terms of shipping and it receives vast quantities of effluents of all types via the many industrialised rivers that drain into it. It is also an important source of food, but overfishing has reduced the stocks of some commercial species to dangerously low levels.

Although some of the more extreme fears about the state of the

North Sea have not been borne out completely, there are a number of problem areas, some of which have been addressed by governments. The marine incineration of toxic wastes has been found to cause localised ecological problems and this is to be halted. Dumped industrial wastes have also caused problems (see Titanium dioxide) and measures have been taken to stop dumping from ships and clean up discharges from pipelines, although there is still room for improvement. Sewage sludge dumping, which also causes localised changes, is to be phased out by 1999. In 1990 high levels of TBTO, a toxic fungicide used as an anti-fouling compound on ships, were found in much of the North Sea. This may pose a serious hazard to marine life.

Despite the fact that some of the more concentrated sources of pollution are now to be phased out, or have been already, there are still problems with materials carried down the major rivers such as the Rhine and the Elbe. Run-off of fertilisers from farmland, together with sewage effluents, contribute a large nutrient load to the North Sea and in some areas eutrophication is occurring. The main problems have been found in shallow coastal areas on the continental side, notably off Norway, Denmark and in the German Bight, although the Thames, Humber, Tyne and Forth are also thought to be at risk.

In March 1990 an intergovernmental conference at The Hague agreed on a reduction of 50% or more in releases of 37 materials into the environment with a view to protecting the North Sea. This included cuts in emissions of dioxins, lead, cadmium and mercury of 70% and also the phasing out of all uses of PCBs by 1999.

NUCLEAR WASTE   While a nuclear reactor operates, some of the fuel – usually uranium – is converted into other materials. These build up in the fuel elements and gradually the fuel becomes less efficient at producing heat. It then needs to be replaced. The spent fuel is removed from the reactor, cooled in water for three months or so and, usually, transported to Sellafield for reprocessing.

Reprocessing generates three groups of material: uranium, which may be reusable as fuel; plutonium, which can be used as a fuel in some reactors and, if of the correct type, as a weapons material; and nuclear waste. The waste contains fission products, formed when nuclei of uranium disintegrate, and actinides (otherwise known as transuranics) which form when uranium absorbs particles called neutrons during the nuclear reactions which power the reactor. The fission products are intensely radioactive and, as a result, produce a lot of heat. The actinides are less intensely radioactive but have much longer lifetimes than the fission products.

This type of nuclear waste – known as high-level waste – is in the form of a solution of materials in nitric acid. It has to be cooled day and night to remove the heat which it generates and the tanks in which it is kept are heavily shielded to reduce the exposure of the workforce to radiation. The radioactivity of the waste will decline comparatively rapidly at first – after some 500 years much of the activity from the fission products will have disappeared – but the waste will remain dangerous for many thousands of years because of the actinide content. Current proposals for dealing with nuclear waste involve converting it into glass blocks and placing these deep in the ground. The technology for doing so has not been perfected and suitable sites have not yet been proven – few places in Britain can be assumed to remain free from earth tremors etc. for tens of thousands of years. The volume of waste for disposal is not great – it would fill a four or five bedroomed house at the moment – but the size of the problem is determined by the amount of radioactivity present and not its physical dimensions.

There are other types of radioactive waste. Intermediate level waste is produced in various processes involving nuclear materials and is not as hazardous as the high-level material. Nevertheless it still needs to be kept out of the environment and suitable places for disposing of it are few and far between. There is a large volume of intermediate-level waste for disposal and even more low-level waste. Low-level waste is produced by hospitals, laboratories and industries as well as nuclear power establishments and can be handled with fairly simple precautions. Some is dumped in landfill and some is incinerated, while a large volume was formerly dumped by Britain at sea. Environmentalist and trades union action caused this practice to be suspended.

The danger of discharging radioactive waste into the environment is that it may return to harm humans or other animals. Some of the radioisotopes involved can bioaccumulate and most of them can be taken up by tissues which use the non-radioactive forms as essential parts of their cellular processes – iodine, for instance, concentrates in the thyroid gland whether it is radioactive or not. Some radioactive elements resemble, but are not identical to, others which are essential; both strontium, which has a radioactive form called strontium-90, and plutonium resemble calcium in some ways and hence are taken into the bones. Once absorbed, radioisotopes can subject vulnerable tissues to radiation and may cause cancer.

Radioactivity differs from some other components of wastes in that its harmful properties do decline with time. However it can take a very long time for this to happen and, at the moment, the rate at which radioactive wastes are accumulating is increasing. Apart from material produced by the nuclear power and weapons programmes there is an unknown quantity of radioactive wastes discarded by industries such as luminous instrument manufacture. This has caused problems in many countries including Britain. A proposed development site on the Essex marshes was recently found to be contaminated with such materials.

OIL   The extraction and transport of oil inevitably leads to leaks of the material into the environment. Major spillages are newsworthy and the images of dying seabirds and polluted coastline from incidents such as those involving the Torrey Canyon, the Amoco Cadiz and the Exxon Valdez are unforgettable. In fact, more oil reaches the sea from spillages on land and the careless disposal of motor oil than from tanker accidents, on average, but this attracts less publicity.

Large amounts of oil in the sea are harmful for several reasons. Seabirds become covered in oil which then poisons them as they try to clean themselves. Layers of oil inhibit photosynthesis in plants and smother coastline ecosystems. Some of the lighter components of crude oil are greenhouse gases and as the oil evaporates they contribute to global warming. Heavier components may be carcinogenic and their presence in food chains may pose a risk to a wide variety of organisms. Layers of oil on rivers and lakes prevent the entry of oxygen from the air.

Dealing with oil spills is difficult, expensive and not always effective. Booms and collection equipment can retrieve spilt material if sea conditions are calm but these are not feasible in conditions such as those on the Alaskan coastline. Control measures often rely on breaking up the oil slick with dispersants in order to increase the rate at which it evaporates or degrades and to prevent it from forming a continuous barrier to birds. This technique is only partially effective and the dispersants themselves can be toxic to wildlife. Fifteen years after the Torrey Canyon disaster, no trace of the oil could be found on Cornish beaches, but some places still carried residues of the dispersants used against the oil.

A more modern approach is to use carefully designed absorbents to soak up the oil which can then be collected and dealt with, although this is not always feasible in bad weather conditions. Biological treatments are also looking promising, although more work on these needs to be done. Polluted beach sand is best collected and disposed of – it usually goes to landfill – while rocky shores may be cleaned with detergents or left for the oil to disperse or degrade naturally.

Providing that the affected area is not too large, oil polluted ecosystems can recover. Sea bird populations may take longer, however, particularly if large numbers have been affected and there are other pressures on them such as abnormally bad weather or food shortages.

ORGANIC   The term organic has two meanings in the environmental context. In the strict chemical sense it refers to chemical compounds of carbon, although a few of these – such as carbon dioxide and carbonates – are not normally considered to be organic. Organic chemicals are produced by living things and are also manufactured from coal and oil (both of which were once living).

The other meaning of organic refers to a method of agriculture which does not involve the use of artificial fertilisers or pesticides. Some organic farmers will use pesticides derived from plants – e.g. derris and quassia – but pest control is achieved mainly by other means such as biological control (q.v.). Compost and other natural manures are used as fertilisers and these provide the nitrates, phosphates and other plant nutrients which conventional farmers obtain from synthetic sources.

OZONE   Ozone is a special type of oxygen, the molecules of which consist of three atoms rather than the usual two. It is highly toxic – Victorian ideas that ozone was healthy were completely misplaced – and even at very low concentrations it attacks the eyes, throat and breathing passages. It is used as a disinfectant for water, as an alternative to chlorination, but as it is unstable it does not leave residues in drinking water.

Ozone in the air at ground level is a health hazard in many cities. It is formed by the action of sunlight on nitrogen oxides and hydrocarbons, most of which are emitted by motor vehicles. It is an ingredient of photochemical smog (q.v.) and accounts for much of the health damage caused by this form of pollution. Ozone also damages trees and other plants, attacks materials such as textiles and rubber, and is a powerful bleach causing fabric dyes to fade.

In the upper atmosphere, however, ozone is positively beneficial. Indeed, it is essential for most current life on earth. A layer of ozone surrounds the planet between 15 and 55 kilometres above the surface and protects life at ground level from the harmful effects of ultraviolet light (q.v.) from the sun. A reduction in the efficacy of this ozone shield will lead to an increase in skin cancers, especially a particularly virulent form called melanoma, as well as genetic damage to plants and some animals. Some damage has already occurred and more is likely as ozone-depleting gases such as CFCs (q.v.) are released at ground level and move upwards. It will be several decades before the extent of the damage from emissions to date becomes known. Meanwhile the chemicals in question are still being released.

PARTICULATES Particulates is the general term given to solid particles emitted to the atmosphere. They range in size from grit (clearly visible to the naked eye) to fume (the particles of which can only be seen under a microscope).

The most obvious environmental effect of particulates is the dirtying of surfaces. Soot stained buildings in urban areas demonstrate the effects of many years' exposure to coal smoke; yet diesel emissions, levels of which are increasing, are three times as soiling as coal smoke. High levels of airborne particulates cause haze and reduce both visibility and the levels of light reaching plants which need it for photosynthesis. The reduced visibility in photochemical smog episodes is largely a result of the formation of particulates.

Particulates are hazardous to health, the magnitude of the risk depending on the size and nature of the particles. Smoke is certainly a lung irritant, especially when combined with sulphur dioxide. Large particles are deposited in the upper parts of the respiratory system, but very fine particles can be carried deep into the most sensitive regions of the lung. If they remain there and consist of hazardous materials such as carcinogenic polycyclic aromatic hydrocarbons, they can expose particularly vulnerable areas to their toxic potential. Some materials, such as very fine particles of lead, are absorbed through the lungs and can be carried in the bloodstream to damage other organs.

Major controls on particulate emissions were achieved by the Clean Air Acts in Britain which have greatly restricted the burning of smoky fuels in urban areas, although in some places, where miners receive free coal, there are still high levels of smoke. Motor vehicles now contribute increasing levels of particulates to urban air either directly, through dirty diesel emissions, brake and tyre wear, or indirectly in photochemical smog episodes. Construction activities, dusty industries and wind-blown soil also add particles to the air.

PCBs PCBs – polychlorinated biphenyls – are highly persistent chlorinated hydrocarbons which have contaminated the environment in many areas of the world. They bioaccumulate and are toxic although it is not certain how much of their toxicity derives from impurities such as dibenzofurans which they normally contain. PCBs are thought to have been responsible for the decline in seal populations in the Dutch Waddensee and over a thousand people were poisoned by them in Japan when they contaminated rice oil. An alarm was raised in the summer of 1990 when a baby dolphin, stranded on the Welsh coast, was found to contain high levels of PCBs as well as DDT. More research into levels of the material in mammals around British coasts is underway.

The manufacture of PCBs has now ceased in the Western world but large amounts remain for disposal. Few completely safe techniques for destroying PCBs have been proven commercially and incineration is the most popular method. When PCBs are incinerated it is vital that temperatures high enough to destroy the material and prevent the formation of dibenzofurans and dioxins are reached – the current standard is 1,100 degrees Celsius with at least 2 seconds residence time in the high-temperature area of the furnace. There must also be sufficient oxygen present. In Britain two incinerators currently burn PCBs and there has been considerable controversy over the importation of PCB-containing wastes for burning. Greenpeace believes that producer countries should arrange disposal, but it has not always been possible for them to do so.

PESTICIDES The term pesticide is a general one, including all chemicals which are designed to kill or control weeds, insects, mites, molluscs, plant diseases and other fungi, rats, mice and any other species which damage crops or otherwise harm or irritate humans, diseases excepted.

Modern agricultural systems and public health programmes are heavily dependent on pesticides, yet at the same time over-use of these materials is harming the environment and people in it. Some specific problems are discussed under Herbicides, Insecticides and Fungicides but a few general points can be made here.

First, pesticides can often harm non-target species such as predators which would normally help to control the populations of a pest or other unrelated organisms. Insecticides have harmed birds of prey and residues of the herbicide paraquat on stubble kill hares. Second, humans can be harmed by contact with pesticides, either as workers using them or as innocent bystanders. The World Health Organisation has estimated that there are some 500,000 to 1,000,000 cases of pesticide poisoning around the world each year, mainly in developing countries, with 5,000 to 10,000 fatalities. In Britain Friends of the Earth have documented hundreds of incidents where people walking or living in the countryside have been made ill by sprays. Third, residues of pesticides contaminate the food chain and also turn up in drinking water. The British Medical Association has concluded that most of the UK population receives low-level exposure to pesticide residues in tap water.

Any new pesticide has to undergo rigorous testing before being released onto the market. There are, however, many old chemicals which were introduced before the current system became operative and which have not been tested so thoroughly. They may present risks to human health and the environment but have been allowed to remain on sale pending further testing, a process which will take decades to complete at current rates of progress. Several older products have been withdrawn in recent years as a result of new evidence of harm coming to light, and there may well be others which should no longer be available.

Most large-scale agricultural systems are highly dependent on pesticides and disease control often requires insecticides to kill vectors such as mosquitoes. Moves towards organic or low-input farming are being made, but they are slow. Thus it is vital that those pesticides which are used should be as safe as possible for the environment and for humans, and their use should be kept to a minimum. This is not the case at present.

PETROL   Petrol is a complex mixture of hydrocarbons with a range of additives designed to improve its performance in vehicle engines. The most hazardous is lead (q.v.) but the use of unleaded petrol is increasing. Petrol vapours are toxic if inhaled in quantity, especially if the fuel contains much benzene (q.v.), and also contribute to photochemical smog (q.v.). The problems caused when petrol is burned are discussed under Motor vehicles.

PHOSPHATES   Phosphates are naturally occurring compounds of the element phosphorus. Small quantities are required by most living cells and they are also important plant nutrients: large quantitites are used as fertilisers. Phosphates are also used as water softeners in most washing powders and have a wide range of other uses, for example as food additives.

When sewage effluents containing large amounts of phosphates are discharged into rivers and lakes they can cause eutrophication (q.v.) and this has been an increasing problem in Britain in recent years with outbreaks of toxic blue-green algae occurring. The situation is complex and phosphates are not the only factors involved: nevertheless they do play a part in eutrophication and algal blooms.

Phosphate run-off from fields may be important in agricultural areas although phosphate is not as mobile as nitrate. Concern about water pollution from phosphates has led to the introduction of a range of phosphate-free washing powders which should help the situation in some areas, such as the Norfolk Broads. Phosphate washing powders are already banned in Switzerland and Canada, because of concern for the health of their lakes. Phosphate removal from sewage effluents is likely to become increasingly necessary since detergents are not the only sources of these substances in sewage.

PHOTOCHEMICAL SMOG see Smog.

PLASTICS Plastics are not generally poisonous in the environment although when they burn they can give off toxic fumes. They may cause a hazard to wildlife, however, and obviously constitute a visually offensive litter problem in many places. Plastics harm wildlife when animals accidentally swallow them or become trapped inside plastic containers.

Plastics are usually very long-lived in the environment, a factor which aggravates any problems which they may cause. In order to alleviate this a range of degradable plastics has been produced. Some rely solely on the action of bacteria – biodegradable plastics – while others (photodegradable types) depend on the action of sunlight triggering a chemical breakdown process which is followed by biodegradation.

Degradable plastics can certainly help with the litter problem and in some other specialised circumstances. The plastic yokes which hold beer cans together have caused problems in harbours by jamming the engines of boats and, in some areas, must now be degradable. However the vast majority of plastics comes from non-renewable fossil fuel sources and it makes more sense to control their use and recycle them where possible than to accelerate their breakdown. In this respect, biodegradable shampoo bottles come a poor second to refillable ones. Environmental groups such as Friends of the Earth generally take a dim view of degradable plastics and would prefer to see plastics used only for purposes where their properties of durability and resistance to degradation are essential.

See Vinyl chloride.

PLUTONIUM Plutonium is a metallic element produced in nuclear reactors. It can be used as a nuclear fuel – the fast breeder type of reactor uses plutonium – and if it consists of a particular isotope it can be used in nuclear bombs. Nagasaki was destroyed by a plutonium bomb.

Although it is not intensely radioactive – a lump of plutonium can be handled with only gloves for protection – plutonium is extremely dangerous. It emits alpha radiation and if a small particle is inhaled into the lung it will irradiate lung tissue intensely, producing a fatal cancer. If plutonium is absorbed in a soluble form it tends to go into the bones where it irradiates the bone marrow. The result can be leukaemia.

Plutonium is discharged from the Sellafield (q.v.) reprocessing plant in liquid effluents and is accumulating in the Irish Sea. Some of this plutonium is blown inland as particles washed onto beaches dry out and are picked up by the wind. Relatively high levels have been found in homes in West Cumbria and, at the time of writing, legal actions alleging health damage by plutonium are pending against the operators of Sellafield.

POLLUTION CONVERSION Pollution conversion is

the process by which solving one pollution problem creates another. For instance, scrubbing power station chimney gases to control acid emissions can result in the production of large quantities of gypsum slurry. If disposed of carelessly, this can cause pollution, so an air pollution problem is converted into a water or land pollution problem. Similarly, burning hazardous wastes without proper pollution control measures can convert a land pollution problem into air pollution. In order to ensure that pollution conversion is minimised, Britain is moving to a system of integrated pollution control (IPC) for a number of industrial activities. IPC is designed to ensure that the environment as a whole is protected rather than one sector at the expense of another. The system is to be introduced gradually and will not be in force fully until 1994 or later.

POLYCHLORINATED BIPHENYLS see PCBs.

POLYCYCLIC AROMATIC HYDROCARBONS

Polycyclic aromatic hydrocarbons (PAHs) are organic chemicals whose molecules consist of several rings of carbon atoms joined together with varying amounts of hydrogen attached. They are emitted by the combustion of many different materials including wood, coal, oil, tobacco and various wastes. Some of them are powerfully carcinogenic. One of the materials in tobacco smoke that has been linked with cancer is a PAH called benz-*a*-pyrene. Although they can be degraded by micro-organisms, levels of PAHs in soil are rising and this has been attributed to emissions from motor vehicles. The cancer-causing potential of these materials seems to be enhanced when they react with ozone, which is also produced as a result of motor vehicle usage.

PVC   see Vinyl chloride.

RADIATION   Radiation is simply a means by which energy is transferred from place to place. The sun's heat and light, the waves in a microwave oven and the signal which carries television programmes are all examples of radiation. Most radiation is electromagnetic (see Chapter 1) and consists of waves – like ripples on a pond when a stone is dropped into it. Some types of nuclear radiation are not waves but consist of fast-moving particles.

Depending on the amount of energy which it carries, radiation may be ionising or non-ionising. Ionising radiation is so called because it is energetic enough to disrupt molecules by converting some of their atoms into ions. Ionising types are generally the most hazardous to life although some non-ionising radiation, such as ultra-violet (q.v.) and microwaves, can be harmful.

The ionising radiations which are of most significance are known as alpha, beta, and gamma although X-rays may also be emitted by some radioisotopes (q.v.). Alpha radiation consists of relatively heavy particles. These do not travel far through tissues or other materials but can do a lot of damage on the way because of their mass. Beta particles are much smaller and faster and can penetrate further. Gamma rays are not particles at all; they resemble X-rays and can penetrate deeply into many materials.

Radiation harms living things in two ways. Simply by delivering large amounts of energy into the body in a short time it can cause burning and serious disruption to the tissues; this is the basis of sunburn and radiation sickness. If the radiation is sufficiently energetic – and ionising radiation is – it can disrupt the genetic mechanism of the cell and cause cancers or mutations. Ultraviolet radiation can also have this effect.

We are surrounded by sources of radiation, from the sun, from the earth's materials and from other living things. In most instances our exposure from artificial sources is much less than from natural sources, although people who receive frequent X-rays will augment their natural dose significantly. Average doses, however, mask the fact that some people may receive much more than others from human sources such as the nuclear programme or accidents such as Chernobyl. As far as we can tell, there is no safe dose of ionising radiation and even background ('natural') radiation appears to be linked with leukaemia. Thus any process which increases our exposure to radiation must be viewed with considerable caution.

RADIOACTIVITY
Radioactivity is the phenomenon in which the heart of an atom – its nucleus – spontaneously changes its nature slightly (decays) and emits a ray or particle of ionising radiation. There is no way of knowing when a particular nucleus is going to decay and no way of influencing the rate at which this occurs. Nevertheless the emission of radiation from a sample of radioactive material is predictable and declines with time. The rate of decay is fastest when there is a large amount of material present and the time taken for half the material to decay radioactively is known as the half-life. Half the remaining material will decay during another half-life, 50% of the remainder in another half-life, and so on. The length of the half-life for a particular radioactive substance is fixed and after ten half-lives it has virtually disappeared. Half-lives can be extremely short – minute fractions of a second – or very long. Intensely radioactive materials tend to have short half-lives while less 'hot' materials may have half-lives of thousands of years. Most of the naturally-occurring radioactive materials in the earth's crust have long half-lives and are not extremely radioactive. Short half-life materials are produced in nuclear reactors and weapons tests.

Radioactivity is a means by which an unstable nucleus attains a stable state. Several steps may be involved but eventually a stable non-radioactive form will be reached.

RADIOISOTOPES
Most chemical elements can exist in two or more forms which have the same chemical properties but which have heavier or lighter nuclei (the nucleus is the core of the atom). These are known as isotopes. Where one of these forms is unstable and decays radioactively it is known as a radioisotope. For example the gas tritium is a radioisotope of the gas hydrogen. It behaves in chemical reactions in the same way as ordinary hydrogen, but has slightly different physical properties in addition to its radioactivity, because a tritium atom weighs about three times as much as an ordinary hydrogen atom.

RADON
Radon is a naturally occurring radioactive gas produced when unstable elements in rocks decompose. It seeps up

through the soil into buildings and may also be produced in the stone from which some buildings are made. The main sites of radon pollution in Britain are in the south-west, with Cornwall having high levels of radon derived from its underlying granite rocks. When inhaled, radon poses a risk of lung cancer since its decay products are solids and remain within the lungs, irradiating sensitive tissues. Uranium miners in poorly ventilated mines where much radon is present have experienced greater rates of lung cancer than comparable populations not exposed to the gas. Some 2,500 cases of lung cancer each year are thought to be caused by radon in British homes.

As many as 100,000 homes in Britain are thought to be affected by excessive radon levels with the major counties concerned being Devon, Cornwall and Somerset. Householders whose homes are identified as radon 'hot spots' may be advised to install ventilation and other measures to reduce their exposure to the gas. It is not completely effective to seal walls and floors against radon as it can permeate many building materials with ease, although cracks and fissures are certainly worth sealing.

SCRUBBERS Scrubbers are devices fitted to industrial plants emitting toxic or corrosive gases. The purpose of a scrubber is to prevent air pollution and it usually works by washing the effluent gases in water or a solution which dissolves the toxic components. Acidic gases, for instance, are absorbed in an alkaline solution such as sodium hydroxide. Various designs exist – the gases may be passed up a column down which the scrubbing liquid flows or the liquid may be sprayed into the gas stream – and each device is tailored to the particular task required. Wet scrubbers also remove particles suspended in the gas stream but other techniques may be required for this such as electrostatic precipitators (q.v.)

SECONDARY POLLUTANTS Secondary pollutants are materials which are formed by the interaction of emissions – primary pollutants – with each other and the environment. For instance, nitrogen oxides and unburnt hydrocarbons are primary pollutants emitted by motor vehicles. These react together with sunlight and oxygen to form the secondary pollutant ozone (q.v.)

SELLAFIELD Situated on the Cumbrian coast, Sellafield has become a byword for pollution from the nuclear industry. It was originally known as Windscale but changed its name following a series of embarrassing incidents in the 1970s and early 1980s. It was as Windscale that it first hit the headlines.

In 1957 a serious fire broke out in a reactor which was making plutonium for the British nuclear weapons programme. Fortunately for English residents (but not so for Ireland) the wind was blowing out to sea for much of the time so a good proportion of the radioactive materials which the fire released was blown offshore. Nevertheless, extensive contamination of farms by a radioisotope of iodine occurred and much milk was destroyed as a result. Little account was taken of other radioisotopes at the time but later work suggested that contamination by a form of polonium, which was also released, probably caused a number of deaths from cancer.

The piles which were making plutonium in 1957 have long been shut down – although the damaged one is not yet decontaminated – but Sellafield continues to operate, reprocessing spent nuclear fuel. After nuclear fuel has been in a reactor for some time it ceases to be effective and has to be removed because fission products (wastes from the energy-producing reactions) accumulate. Another material, plutonium, is also created in the fuel. Reprocessing separates the fission products and the plutonium from the fuel which can then be reused. The fission products are, by and large, useless but plutonium can be used in weapons or in some types of reactor. The fission products, and a group of materials similar to plutonium called actinides, constitute nuclear waste (q.v.)

Reprocessing is a messy business and over the years large amounts of radioactive material have been discharged to the environment. It is estimated that the Irish Sea contains about a quarter of a tonne of plutonium from Sellafield and traces have been found around most of the British coast. In the early 1980s a 20-mile stretch of beach was closed because of contamination from the plant and some years previously a leak of material from a sampling point the operators had forgotten about contaminated a building. Prior to that Windscale – as it then was – became the focus of attention when a tank containing low-level radioactive waste was found to be leaking, an event which was concealed from the Secretary of State for Energy, who was responsible for the nuclear industry. Many other incidents have occurred and information about them is by no means complete.

There have been many calls for the closure of Sellafield but these have been resisted for several reasons. First, the plant makes nuclear weapons material for the Ministry of Defence. Second, the civil side of its operations includes some lucrative contracts to reprocess waste from abroad, notably Japan. Third, some of the spent fuel from

Britain's reactors is so badly corroded that it must be reprocessed quickly or it will disintegrate and leak radioactive materials.

There are alternatives to reprocessing some types of spent fuel but it is certain that Sellafield will be operating for many years to come.

SEVESO   In 1976 a chemical factory at Seveso, near Milan, suffered a major accident when a manufacturing process ran out of control, bursting a pressure release device and spewing toxic material over the surrounding neighbourhood. The plant was making trichlorophenol, a substance which is used to make the herbicide 2,4,5–T and the antiseptic hexachlorophane. An inevitable contaminant of trichlorophenol is a form of dioxin known as TCDD and during the runaway reaction much more of this material was produced than is normally the case.

The immediate effects of the leak on the environment included the deaths of wildlife, cats and dogs. Some 70,000 animals died or were put down as a result. Humans suffered as the painful skin condition chloracne (q.v.) appeared in exposed people. Up to 500 people were estimated to have been affected by one form of illness or another.

The long-term effects of the accident are unclear. An increased rate of birth defects in babies born to mothers exposed to TCDD might have been expected, but health statistics in the area were unreliable, so no-one is sure what the 'normal' level of defects is and comparisons cannot be made. Furthermore, ninety pregnant women in the area had abortions as they feared that their babies might be deformed. The miscarriage rate was reported to have increased significantly just after the accident.

Soil, crops, roads and buildings were contaminated with TCDD and remained so for many years. Some of the most badly polluted material was excavated and put into drums for disposal. These eventually turned up in a French slaughterhouse where they had been stored illegally. In 1985 this material was finally burned in a Swiss incinerator.

Although the health effects of the accident are still not established completely, it did lead to a revision in legislation concerning major chemical hazards. An EC Directive, known as the Seveso Directive, now requires improved monitoring, control and emergency planning for plants likely to cause serious harm to the community in the event of an accident.

SEWAGE   Something like 97% of sewage is water: it is the remaining 3% which can cause serious pollution and hazards to health if incorrectly handled. Most homes in Britain are now on mains sewerage networks, although in remote areas septic tanks and soakaways are still used to dispose of such wastes. Sewers generally lead to some form of treatment works although in many coastal locations untreated sewage is still pumped out to sea.

The purpose of sewage treatment, where it is undertaken, is to remove or at least reduce the polluting potential of the sewage and also to minimise any disease risk. The polluting potential derives from the organic matter and plant nutrients present, although many sewage flows are also contaminated with effluents from industry which may contain heavy metals or other toxic substances. Bacteria and viruses constitute most of the disease risk in sewage.

The first stages of sewage treatment involve the removal of solid matter which can be anything from lumps of wood to particles of vegetable matter and sand. Screening removes the larger items while settling tanks permit smaller particles to fall to the bottom. The remaining material, which consists of water with dissolved materials and some suspended matter, is then treated biologically. Traditional processes involve trickling the liquid over a bed of coke where populations of micro-organisms use the organic matter as food. In modern activated sludge plants the micro-organisms, in the form of sludge, are stirred into the sewage and the mixture is aerated and agitated. These processes reduce the BOD (q.v.) of the sewage dramatically and also convert ammonia into nitrate. Over 99% of the pathogenic bacteria are destroyed as well. Before the treated effluent can be discharged it has to be allowed to settle so that much of the remaining suspended material settles out. Further treatment may be necessary to remove nitrates or phosphates in order to prevent eutrophication in the receiving waters.

Both processes described above produce large volumes of sludge which has a high BOD and may be contaminated with heavy metals. This can be digested to produce methane, a useful fuel, but much material will still remain for disposal. Currently, sea dumping is a popular route for disposal but this is being stopped by 1999 on environmental grounds. Some sludge is spread on the land as a fertiliser and this practice may be increased, subject to the limits set by the metals content. Some is landfilled and some is incinerated but neither of these options is as cheap as dumping at sea.

Sewage discharge with little or no treatment on or near beaches causes considerable problems in coastal areas. The problems are aesthetic – bathers are understandably repelled by faeces and condoms floating past – and microbiological. Various minor infections, such as stomach upsets and earaches, are attributed to exposure to sewage-contaminated water and if the sewage contains any really dangerous organisms, such as typhoid bacilli, a serious hazard could occur. Current tests for microbiological purity – which many British beaches fail – look only at one type of bacterium (faecal coliforms) as an indicator of pollution. They do not investigate other bacteria such as *Salmonella* or viruses which can also cause disease. Some medical authorities believe that more extensive testing, for pathogenic organisms, should be carried out before a beach can be declared safe for bathing.

Rivers are also polluted by sewage. Many of Britain's sewage works are old and in need of repair. Many are overloaded and now serve a population much greater than that for which they were originally designed. As a result, standards for discharges are not met and gross pollution of rivers results. The standards themselves were not particularly tight originally, and many were relaxed in the run-up to water privatisation.

Until 1989 the water authorities who ran the sewage works were responsible for monitoring and enforcing standards on discharges, but this is now the job of the National Rivers Authority (NRA). Unfortunately, during its first twelve months of operation, although it took many samples of effluents which failed consent conditions, it was unable to prosecute any of the water authorities since the samples taken were not the required three-fold samples. Public registers of analytical results showed that 392 sewage works breached content conditions during the NRA's first year.

SICK BUILDING SYNDROME   see Indoor air pollution.

SMOG   The term smog has two meanings in the context of air pollution. The classic 'London smog' was a combination of smoke and fog which formed when the emissions from coal-burning combined with fog and were trapped over London by weather conditions for several days. The 1952 smog caused the deaths of some 4,000 people, mainly by aggravating existing heart and lung diseases. Smoke particles caused lung irritation, compounded by the presence of sulphur dioxide which dissolved in the water droplets in the fog to form acids. Other cities around the world have suffered from this type of smog, and there have been many fatalities, but controls on coal burning in urban areas effectively prevent this form of pollution.

The other type of smog, photochemical smog, results from pollution from motor vehicles and is a growing problem worldwide.

First documented in Los Angeles, photochemical smog is now a serious health hazard in cities as diverse as Sydney, Athens and Jerusalem. It is also an emerging problem in Britain with high levels of pollutants recorded in London, Birmingham, Oxford and Bristol. Nitrogen oxides, produced when fuel burns in vehicle engines, interact with sunlight and the oxygen of the air to produce ozone. This further reacts with nitrogen oxides and other components of vehicle exhausts to produce a brown smog which irritates the eyes, throat and respiratory passages. The ozone levels which result can cause serious harm to people with lung and heart problems, particularly the elderly, and even school pupils are advised to refrain from physical exertion when smog warnings are issued in California. Eye irritation is caused by compounds called PANs which form from ozone and unburnt hydrocarbons.

As with earlier smogs, photochemical smog forms when temperature inversions and other climatic phenomena trap the vehicle emissions in a confined sunny area. Other sources of volatile organic compounds, such as paint solvents and even natural oils, can also contribute to the formation of photochemical smog.

SOLVENTS   Solvents are simply liquids which dissolve other materials. Water is the commonest example. A more specific use of the term, however, relates to organic solvents which are used in a range of products from glues to cleaning fluids. Most solvents are volatile organic compounds (q.v.) and many, such as toluene and xylene, are hydrocarbons. Some, such as trichloroethylene, are chlorinated hydrocarbons which pose particular problems if misused.

Air pollution may result from the use of hydrocarbon solvents, as they can form photochemical smog, while the careless disposal of chlorinated solvents has contaminated groundwater (q.v.). Exposure to some of these materials at work or in the home can be dangerous; some are linked with birth defects while others can produce intoxication. Users of certain adhesives and wax polishes sometimes feel light-headed while the deliberate abuse of solvents, when large quantities are inhaled in order to produce euphoria, claims several lives each year.

Some of the more toxic solvents are now controlled. Carbon tetrachloride, which used to be used for removing grease spots from clothing, is no longer available to the general public and is being phased out industrially; it is also a greenhouse gas. Low-solvent paints are coming onto the market and controls may be imposed on products containing volatile organic compounds on air pollution grounds. The problem of solvent contamination of groundwater remains, however, and this may increase as more of these materials are discarded with domestic rubbish.

SULPHUR DIOXIDE
Sulphur dioxide is a colourless toxic gas which is formed whenever materials containing the element sulphur are burned. It dissolves in water to produce sulphurous acid, which combines with oxygen to form the stronger sulphuric acid. Sulphur dioxide itself also reacts with oxygen to form sulphur trioxide, which dissolves in water to form sulphuric acid.

Sulphur dioxide is, generally, the main compound responsible for acid rain (q.v.). Other gases play a part either in contributing acidity directly or in bringing about the conversion of sulphur dioxide to sulphuric acid. The old-style London smogs contained large amounts of sulphur dioxide from burning coal and this was responsible for part of their toxic impact, since sulphur dioxide is a lung irritant, particularly in conjunction with smoke.

The gas can be removed from chimney emissions by a variety of scrubbing devices. Most work on the principle of reacting the gas with limestone, a process which produces gypsum as a byproduct. This may be usable for making plasterboard but may also be dumped in landfills where it can cause pollution. An alternative process, the Wellman-Lord process, generates other useful materials as a byproduct and involves less limestone – an important consideration since much limestone comes from national parks. Sulphur dioxide can also be removed at the point of combustion if the fuel is burned in a fluidised bed system. Powdered limestone is added to the combustion zone and the sulphur is trapped, forming a solid waste. This can be more effective than the wet scrubbing system but still uses limestone and produces a solid waste.

SUSPENDED SOLIDS
The polluting components of an effluent discharged into a stream may be dissolved in the water but are often in the form of suspended solids – particles of material carried along by the moving water. Organic matter in sewage is an obvious example; much of the BOD (q.v.) of a sewage effluent is associated with the solids content.

Apart from the BOD implications, suspended solids can harm river life in two main ways. Fish and other organisms can be damaged as their gills or other breathing apparatus become clogged by solid matter. Where solids are present at high levels, they make the water cloudy (turbid) and this prevents the penetration of light to plants which require it for photosynthesis. In some cases, such as streams taking china clay washings in Cornwall, the river runs white with suspended solids and photosynthesis is impossible.

Suspended solids are generally removed or reduced at sewage works by allowing them to settle out in tanks through which the effluent moves slowly. Some solids, notably the very fine particles, still emerge in the outflow but good design should reduce these to an acceptable level.

TCDD　　See Dioxins and Dibenzofurans.

TEMPERATURE INVERSIONS　　The normal pattern of temperature in the atmosphere is for the air to become progressively cooler with increasing height above the ground. This means that warm gases – e.g. from chimneys – rise easily and can disperse readily. Under some conditions, however, a lid of warm air can become established over an area and prevent the rise of effluent gases, trapping them beneath the lid. As a result, they can react together to form other pollutants and levels of air pollution will increase until the inversion conditions break down. This phenomenon is a contributory factor to the formation of smogs in cities such as Los Angeles, Athens and London.

TITANIUM DIOXIDE　　Titanium dioxide is a bright white powder which is used in many products to achieve whiteness. Toothpaste, plastics, paints and paper all contain titanium dioxide which is non-toxic and stable. Although the material itself does not cause pollution, waste produced during its manufacture does. Greenpeace has campaigned for many years – successfully in many cases – against the dumping of titanium dioxide manufacturing wastes at sea.

The wastes in question are highly acidic and contain large amounts of dissolved iron. When dumped at sea the acid can burn fish and the iron comes out of solution and forms a thick precipitate which can clog the gills of fish. A wide range of abnormalities in fish have been found in the regions where these wastes have been dumped. As the iron comes out of solution it takes oxygen from the water – the iron in solution is oxidised. This can cause the formation of zones around the dumping area where oxygen levels are low and aquatic life is harmed.

As a result of pressure from environmentalists and, latterly, legislation several European companies have cleaned up their processes either by recycling the acid or treating the effluent. Marine dumping has all but ceased although two companies in Britain discharge wastes through pipelines into the North Sea. Improvements in these discharges have been made and more are in prospect. Despite these improvements, titanium dioxide manufacture is an inherently messy process and Greenpeace advocates a minimisation of the use of the material.

TOBACCO   The health effects of tobacco smoking on users are well documented – increased rates of heart and lung disease, cancers and circulatory problems – and need not be discussed here.

The effects of the material are not confined to smokers, however, since tobacco smoke is a major air pollutant in many homes, workplaces and public buildings. Recent research has shown that passive smokers – those who inhale other people's smoke – run an increased risk of contracting lung cancer since the polycyclic hydrocarbons which are believed to cause the disease are present in the smoke whether or not it is inhaled by the smoker. Furthermore, children in households where one or more adults smoke tend to suffer more from lung problems in their early years. Babies born to smoking mothers are often disadvantaged. For example, they may be premature and of lower birth weight.

Apart from the carcinogenic hydrocarbons, tobacco smoke contains a wide range of other harmful materials. Nicotine – the active and addictive component of the smoke – is obviously present and so is cadmium (q.v.). Carbon monoxide (q.v.) is there in varying quantities but is not normally a hazard to passive smokers, unless they happen to be unborn babies of a smoking mother. If one treats tobacco smoke as a pollutant, and includes smokers as victims of pollution, then smoking kills far more people per annum than all other forms of air pollution put together.

ULTRAVIOLET LIGHT   See UV light.

URANIUM   Uranium is a heavy metal used as a nuclear fuel. Specialised forms of it, consisting of very high proportions of the isotope uranium-235, are also used in nuclear weapons. Uranium is radioactive but not intensely so; it can be handled with only moderate protection against exposure to radiation.

Mining uranium is much more hazardous, however, since the mines usually contain large quantities of uranium decay products such as radon (q.v.). The rubbish left behind when uranium has been extracted from ores – the tailings – is also radioactive and has caused serious health problems in the US when carelessly dumped. Schools and communities have been built on or near tailings dumps and people have been exposed to high levels of radiation as a result.

UV LIGHT   As well as the visible light which enables us to see, the sun emits a wide range of other radiations some of which are harmful. One such type of radiation is ultra-violet (UV), which is responsible for the development of suntans in people exposed to the outdoor sun. Ultra-violet light of a particular type, known as UV-B, is also responsible for the induction of various types of skin cancer of which one kind, malignant melanoma, is often fatal. The skin also ages quicker when subject to ultra-violet exposure. Ultra-violet light also increases the rate of mutations in organisms and is harmful to the small floating plants that form the basis of oceanic food chains.

A thin layer of ozone (q.v.) in the upper atmosphere protects us against excessive exposure to ultraviolet. Before this existed, life was confined to the seas and lakes deep enough for the radiation to be absorbed by water. Now, human activities are damaging this ozone layer and increases in the rates of the various types of skin cancer are expected. Even without damage to the ozone layer, many medical authorities are advising people, especially children and those with fair skins to avoid unprotected exposure to the sun in order to reduce the risk of skin cancer.

VINYL CHLORIDE   Vinyl chloride, more fully known as vinyl chloride monomer or VCM, is the building brick from which the plastic PVC is constructed. It is a gas which is heated under pressure to form PVC. VCM was found to cause a rare form of liver cancer in workers making PVC and industrial hygiene standards have been tightened up as a result. It was also found to migrate into food from some types of plastic wrappings and safer, non-PVC versions are now widely on sale. Some PVC flooring materials exude VCM for a while after they are laid and good ventilation is necessary to prevent the build-up of hazardous levels. These problems result from the fact that PVC often contains mobile traces of VCM which have not been converted to the plastic.

The combustion of PVC results in the production of hydrogen chloride gas, a material which is toxic, corrosive and can shorten the life of chimney linings in incinerators. It is also thought to contribute to the formation of dioxins when domestic wastes and plastic coffins are burned, but there is no simple relationship between the amount of PVC burned and the amount of dioxins produced.

VOLATILE ORGANIC COMPOUNDS Volatile Organic Compounds (VOCs) is a general term given to solvents, gases and liquid fuels which evaporate easily at normal temperatures. Many are hydrocarbons and some are chlorinated hydrocarbons, though other types of chemical are included as well. These materials are of significance since high levels can cause air pollution. Many are involved in the production of photochemical smog and some are also greenhouse gases. Some chlorinated materials may also threaten the ozone layer.

Currently, there are no regulations covering the emissions of VOCs in Britain, although some manufacturers have taken steps to reduce levels from their plants. 3M, for instance, has installed solvent vapour collection equipment in one plant, while ICI has developed a range of low-solvent and water-based paints for cars. Several manufacturers now offer low-solvent paints for domestic use as well. In the US, some states are considering imposing regulations on emissions of VOCs and several countries require cars to be fitted with devices which trap petrol vapours escaping from fuel tanks.

WATER FILTERS In an attempt to remove undesirable materials from drinking water, many people are turning to water filters. This trend has been encouraged by manufacturers and suppliers some of whose publicity material has been less than honest.

It is possible to remove all but very small traces of pesticides, lead, nitrates, aluminium, chlorine and solvents from water but to do so requires the expenditure of hundreds of pounds – and is probably not necessary anyway. Much smaller reductions in the concentrations of these materials can be achieved less expensively, although purchasers of such equipment should insist on seeing details of what the filter will actually do. Some types of filter – such as certain in-line devices fitted to the water supply – may actually increase the health risk from drinking water by allowing bacteria to breed, and water companies can advise on the suitability of particular devices.

Irrespective of any other contaminants, some people find that a water filter improves the taste of drinking water – partly because it removes chlorine and also because it removes some minerals which are not harmful but merely add taste. This is often a matter of personal preference and it must be remembered that a jug of water left standing for long periods in a warm place will tend to grow bacteria, particularly if residual chlorine has been removed.

ZINC Zinc is a heavy metal which is needed for a variety of processes in plants and animals. Indeed, there is evidence to suggest that significant numbers of people may be deficient in zinc as a result of modern diets and lifestyles. In excessive quantities, however, zinc is toxic, although it is much less poisonous than lead, cadmium or mercury. High levels are harmful to plants.

Concentrations of zinc approaching those which inhibit photosynthesis in algae have been found in some areas of the Bristol Channel.

PART 3

FIGHTING BACK

4: WHO PROTECTS US AND HOW

Laws to control pollution in Britain have existed for a very long time. As far back as the thirteenth century, causing air pollution by burning coal was an offence in London but, like many pollution laws since, this was infrequently enforced. Air pollution from industrial sources began to come under control in 1863 with the first of the Alkali Acts, although there had already been some attempts to deal with the problem, often on a local basis. This Act was introduced as a result of pressure brought by wealthy landowners whose livestock and crops were harmed by emissions from the alkali industry in the north of England. The Alkali Acts, much amended, remained the mainstay of industrial air pollution control for well over a hundred years.

The control of non-industrial air pollution was prompted by the great London smog of 1952 which killed four thousand people. Four years later, the first Clean Air Act enabled local authorities to ban the burning of smoky fuels in towns and enormous improvements in air quality resulted. Smoke control areas have now been established in most British towns but in some coal mining areas the burning of coal and other non-smokeless fuels is still permitted.

The control of river pollution lagged behind air pollution control for many years, the main Act not appearing until 1951, although minor and patchily enforced legislation had existed earlier. The 1951 law was the Rivers (Prevention of Pollution) Act and, together with a second act ten years later, set up a system by which consents to discharge effluents were granted by the then river authorities. Prior to that Act, only those who owned the banks alongside a polluted river could take action against a polluter and then only in the civil courts. Discharges of effluents to public sewers were regulated under the Public Health Acts, the main relevant one appearing in 1936.

Radioactive materials were first brought under comprehensive control by a licensing system set up under the Radioactive Substances Act 1960. This remains the mainstay of control, although other legislation has had an impact. Anyone storing, using or discharging a radioactive substance as defined under the Act must be licensed and strict limits are set on the quantities of materials involved.

The 1970s saw the beginning of broader and, theoretically, tighter controls on pollution. The Health and Safety at Work Act of 1974 laid the basis of control of hazardous materials within factories and this obviously had implications for air pollution leaving the plant. More radical was the Control of Pollution Act of 1974 which set out a structure for controlling air and water pollution, noise, and the disposal of wastes. Unfortunately, this provided only a framework and some of the regulations which were supposed to put its principles into practice never appeared. Other 1970s legislation included controls on some emissions from motor vehicles, the licensing of waste dumping at sea and an emergency Act to control the dumping of toxic waste, prompted by the discovery of cyanide on a children's play area in the West Midlands.

During the 1980s few new measures were enacted but some regulations under the Control of Pollution Act were produced and brought into force. The Food and Environment Protection Act of 1985 brought in the first statutory controls on pesticides (hitherto, most controls had been voluntary 'gentlemen's agreements', although not all the parties concerned behaved like 'gentlemen') and measures to phase out the use of lead in petrol were introduced. Water privatisation measures transferred the control of river pollution from the water undertakings to the National Rivers Authority.

Pollution control was overhauled again with the passing of the 1990 Environmental Protection Act. This introduced the concept of integrated pollution control (IPC) and is designed to ensure that all environmental impacts from a scheduled process are controlled, thereby preventing pollution conversion. The Act covers air pollution, waste management, the importation and use of prescribed substances, dumping of waste at sea, stubble burning and litter.

The philosophy behind pollution control in Britain has changed over the years but it still differs from that in other countries. With a few exceptions, the UK has not set strict, generally applicable limits for pollutants at the point of discharge. Industries emitting air pollutants were required to use the best practicable means (BPM) to prevent such emissions and to render them harmless. If a plant could be shown to be complying with BPM then it was legal, irrespective of any harm caused. The term BPM provided considerable scope for negotiation between the firm and the Alkali Inspectorate which enforced the law, the result of which was sometimes lax standards. It did, however, mean that improvements could be required if a new means of pollution control became available. The Alkali Acts included one statutory emission limit, for hydrogen chloride, but later amendments introduced a series of presumptive limits for other processes. These were not legal limits, but if a plant complied with them it was deemed to be operating BPM.

BPM gradually gave way to BPEO, which stands for Best Practicable Environmental Option. This recognised the fact that solving one problem could lead to another – pollution conversion (q.v.) could occur. This, in turn, has given way to the concept of Integrated Pollution Control coupled with BATNEEC – Best Available Technology Not Entailing Excessive Cost. Exactly how this will work out in practice, and how 'excessive cost' will be interpreted, remains to be seen. It should become apparent when the Environmental Protection Act has been operating for a while.

Britain's reluctance to set fixed emission limits – e.g. no more than so much of a pollutant per day from every factory using a given process – has brought it into conflict with the EC which prefers such limits. Britain has advocated the concept of an Environmental Quality Objective (EQO) whereby discharges of a pollutant are regulated in order to meet a particular overall goal for a sector of the environment. For instance, if the EQO for a stretch of river is that it should support coarse fish, all discharges into that river are controlled so as to improve the river to that level. This may mean tighter or more lax controls compared with a fixed standard, depending on local circumstances. The problem has been that when EQOs have been set for rivers they tend to be set at or near the status quo to avoid excessive expenditure by the water authorities – who had to set the standards and also caused the most polllution.

EC legislation has had a profound impact on British environmental law. Many aspects of UK legislation have been prompted, sometimes reluctantly, by Europe under the principle that controls should be uniform across the community to prevent unfair trade advantages. There are two types of EC legislation which are particularly relevant here: Directives and Regulations. A Directive sets a particular goal – the reduction of sulphur emissions, for instance – and allows member states to produce their own legislation to attain this goal. The exact form of the law and the measures taken will vary from country to country but must meet with EC approval. A Regulation is much more specific, requiring certain actions or standards across the community without exception unless specifically provided for.

Most environmental legislation takes the form of Directives and in some instances a 'framework directive' may be produced within which a number of 'daughter directives' subsequently appear. An example of this is the framework directive on dangerous substances discharged into the aquatic environment which listed a range of substances to be controlled. Subsequent daughter directives set limits for particular materials such as heavy metals and pesticides.

Should a member state fail to comply with an EC directive or regulation it may be taken to the European Court. This has happened to Britain on several occasions; Friends of the Earth have prosecuted the British government over drinking water quality, and Britain is not the only defaulter.

The general background to pollution law has been described above. How it applies to particular forms of pollution is outlined below. It must be remembered that the law is changing all the time and the fact that a law exists does not mean that it will be enforced properly or, indeed, at all. The first year of operation of the National Rivers Authority should have led to a number of prosecutions of polluting water undertakings whose sewage works breached consent conditions. This did not happen because the NRA only collected single effluent samples rather than the triple samples required to bring the cases to court. Most of the legislation referred to applies to England and Wales. Similar provisions will normally apply in Scotland, although the enforcing body may be different.

Air pollution

The control of air pollution has been a divided responsibility for many years. Large industrial plants operating 'scheduled processes' were formerly under the control of the Alkali Inspectorate, which took its name from the original act which established it. This is now Her Majesty's Inspectorate of Pollution (HMIP), a body which has wider responsibilities than industrial air pollution control. Other forms of air pollution have been the responsibility of local authority environmental health departments.

Processes causing air pollution, called prescribed processes, are now divided into two groups: Schedule A processes and Schedule B processes. Anyone wishing to operate a prescribed process will need to obtain authorisation and in the case of the first group, this will come from HMIP. The Schedule A processes are the major industrial activities such as smelting, waste incineration, and power generation, and HMIP will be enforcing the concept of Integrated Pollution Control.

Environmental Health Departments will have the responsibility of issuing authorisations for, and monitoring, the Schedule B processes of which there are expected to be about seventy. These processes are individually smaller polluters than Schedule A processes but, nevertheless, their effects on the local environment can be highly significant. Integrated Pollution Control will not be applied to these processes. In addition to these activities, Environmental Health Departments are responsible for enforcing the Clean Air Acts by ensuring that smoky fuels are not used in domestic premises.

Inspectors enforcing the law will have fairly wide powers of entry to premises and will be able to demand documents, take samples and even close down processes immediately if a serious risk exists. During the passage of the Environmental Protection Bill through Parliament concern was expressed that there would be insufficient resources for Pollution Inspectors and Environmental Health Officers to carry out their duties of inspection and enforcement adequately and this had not been resolved at the time of writing.

Motor vehicle emissions

Controls on motor vehicle emissions are based mainly on a system of type approval – a standard is set for emissions from a particular type of vehicle and all new models must comply with it. The standards, which are set by the EC, now prescribe limits for the major polluting gases and are in the process of being tightened up, albeit slowly. In order to meet these standards some types of vehicle will require catalytic converters while others may be able to comply using lean-burn engines, which pollute less than conventional versions, at least for the time being.

British road traffic law has included some provision for pollution control. It is an offence to use a vehicle which emits materials likely to damage property or cause injury to people and any pollution control equipment (including silencers) must be properly maintained. There are also controls on the emission of smoke from diesel engines in use. The enforcing authority is normally the police or the Department of

Transport. This is not an area to which the police have given priority although the DoT has a number of inspectors who check diesel emissions. In 1990 the government promised to introduce a limit for carbon monoxide emissions as part of the annual MOT test for cars.

The composition of motor fuel has been regulated in order to control pollution. There are limits on the amount of sulphur permitted in diesel fuel. The lead content of petrol is also controlled. Leaded petrol is not illegal but all new vehicles must be able to run on unleaded fuel.

Air quality standards

The EC has introduced several Directives which set standards for air quality within the community. These cover sulphur dioxide, suspended particulates, lead in air and nitrogen dioxide. Member states have some latitude as to how they comply with these standards but they are expected to meet them. Currently, several areas in Britain do not meet the limits for particulates (smoke) and have been granted exemptions until April 1993.

River and groundwater pollution

The control of river pollution has long been operated through a system of consents to discharge effluents. Before privatisation the Water Authority granted dischargers – which were often its own sewage works – a consent to discharge a given quantity of effluent with specified characteristics such as BOD, suspended solids, heavy metals content and so on. If these consent conditions were breached, the Authority could prosecute the discharger. Making an unauthorised discharge into a river was also an offence.

This system is retained in its essential aspects but the enforcing authority is now the National Rivers Authority which is independent of the water companies. It can bring prosecutions on the basis of its own sampling programmes although each sample must be taken in triplicate with one being supplied to the discharger for analysis, one being analysed by the NRA and one being kept in store in case of dispute. A register of consents to discharge effluents and the results of analyses taken by the NRA is maintained and is open to inspection by the public. The NRA is also responsible for controlling pollution of groundwater and lakes.

Coastal and sea pollution

The control of pollution at sea is split between several agencies. The dumping of wastes (soon to be halted) is the responsibility of the Ministry of Agriculture Fisheries and Foods, while oil pollution is dealt with by the Department of Transport. Discharges to coastal waters are regulated by the National Rivers Authority.

Drinking water

A good proportion of the NRA's work involves the protection of drinking water since rivers from which water is abstracted for public supply, as well as groundwater, must meet EC standards. The quality of water supplied to households must also meet EC criteria which are embodied in British regulations. The enforcement of these standards is the responsibility of the Drinking Water Inspectorate, although Environmental Health Departments often have a role to play as well.

Hazardous waste

The Environmental Protection Act reinforced the controls on the handling and disposal of waste which were initiated by the Control of Pollution Act 1974. It has set up a system of Waste Regulation Authorities (WRAs), based within county or borough councils, who are responsible for licensing waste management sites in their areas. In the past, some local authorities both operated disposal sites and enforced standards; this has now changed, since the waste disposal side of their activities have been split off from the enforcement sides as semi-independent organisations.

Anyone who wishes to operate a waste management site, be it a tip, an incinerator or a recycling plant, must obtain a licence from the local WRA. Before granting a licence the WRA must be convinced that the applicant is a 'fit and proper' person to hold a licence and has sufficient financial resources to operate the site properly. Conditions as to the types and quantities of wastes to be handled at the site will be imposed by the WRA, which is also responsible for inspecting operations and ensuring that the operator complies with the conditions. HMIP will supervise the work of the WRA, providing another layer of enforcement, and will be responsible for authorising all hazardous waste incineraton plants, which are Schedule A processes. Planning permission must be obtained before a site can be developed. A public register of licences is kept.

A new development is the concept of the 'Duty of Care', another result of EC influence. Anyone producing or handling a waste has a responsibility to ensure that it is dealt with in a proper manner and this duty is indefinite – someone will always remain responsible. A landfill site is the responsibility of the operator until all risks of gas generation or pollution of water have disappeared, which can take decades after the site is closed.

Special controls exist on the most dangerous types of wastes, known as 'Special Wastes'. A consignment note system is designed to ensure that the local authorities in whose area the waste originates and the authority in whose area the waste is to be dealt with are informed of any such waste shipment. The transporter must also be fully informed about the nature of the waste being carried. Detailed records must be kept of the fate of the waste. The international transportation of waste is also controlled and the Secretary of State for the Environment has the power to prevent the importation of particular consignments of waste where necessary. The Secretary of State can also direct that a given waste be dealt with at a particular site.

Noise

Noise within industrial premises is the concern of the Health and Safety Executive whose Factory Inspectors are responsible for enforcing standards designed to protect employees. Outside factories, however, Environmental Health Departments have the major role and they spend a large amount of time dealing with complaints about noise from domestic premises and other sources in the community. Most of their powers stem from the Control of Pollution Act 1974 and they have powers to serve abatement notices requiring the noise to be stopped. In some circumstances an aggrieved citizen can bring a private action for noise nuisance in the civil courts, but this can be expensive.

Planning controls can be used to prevent noise nuisance by siting noisy developments away from residential buildings. The preventive approach is also adopted in regulations limiting the noise from motor vehicles and other mechanical sources such as construction plant and lawnmowers. The police can prosecute motorists driving vehicles with defective silencers, although prosecutions are uncommon. People affected by noise from new road developments may be entitled to compensation if the sound levels are high enough, but this does not apply when traffic suddenly increases on an existing road.

Pesticides

Until fairly recently, controls on pesticides were mainly voluntary, but the Food and Environment Protection Act of 1985 introduced a statutory basis for controlling all types of pest control compounds. Any material marketed as a pesticide must be registered and approved by a government committee responsible to the Ministry of Agriculture. Information must be supplied by the manufacturer as to the product's efficacy, safety and behaviour in the environment. Regulations prescribe the crops or pests on which the product can be used and the warnings and instructions which must appear on the label. Anyone selling pesticides to the public or to professionals must be adequately trained to give appropriate advice with professional sales staff, as opposed to shop assistants, being required to hold a certificate of competence.

The users of pesticides are required to use the materials safely and in accordance with the instructions on the label. In particular, users must take 'all reasonable precautions to protect the health of human beings, creatures and plants, to safeguard the environment and in particular to avoid pollution of water'. This basically entails compliance with the official code of practice published by the Ministry of Agriculture; although it is not an offence not to follow the code, if anything untoward happens failure to comply can be used in evidence against the operator. Under the Control of Pesticides Regulations 1986, made under the Act, operators must have a certificate of competence to use pesticides and a similar requirement is placed on storekeepers holding the materials.

Enforcement of the pesticides regulations is the responsibility of the Agricultural Inspectorate, a small body which has limited resources in the field. Pesticide contamination of rivers and drinking water is the concern of the National Rivers Authority, while Trading Standards Officers and, sometimes, Environmental Health Officers are interested in pesticide residues in food. Aerial crop spraying is regulated by the Civil Aviation Authority, but accidental spraying incidents should be reported to the Health and Safety Executive.

Major hazards

Following the Seveso (q.v.) incident in 1976 the EC drew up a Directive to control major hazards – installations which could cause serious damage to the surrounding community in the event of an accident. This has been embodied in regulations made under the Health and Safety at Work Act in Britain and such plants, which are defined in the regulations, must now draw up emergency plans to cope with accidents. The plant operators must also notify the local population of measures to be taken in an emergency – e.g. a major escape of toxic gas – and draw up evacuation plans if appropriate. New regulations now apply to the storage of large quantities of hazardous chemicals, as well as to their production or use in manufacturing.

Toxic substances

There is no overall piece of legislation controlling toxic substances, although many laws and regulations made under different acts are relevant. Some of these, such as the Control of Pesticides Regulations, have been described already and others – such as the Poisons Act – are not directly relevant to the environment. Within the workplace, the Control of Substances Hazardous to Health Regulations require much more information on a material's hazards to be made available to employees than hitherto. The safety data sheets provided under these regulations can provide useful information on the toxicity of a substance which may help in assessing its potential harm to the environment.

All new substances to be manufactured or used in quantities of one tonne per year or more are now subject to regulations which require manufacturers or importers to provide information to the Health and Safety Commission about the substance. This information must include details of toxicity and also some data on its properties in the environment. The Secretary of State for the Environment has the power to ban the importation or use of any given substance on the grounds of potential environmental damage and some substances – such as blue asbestos – have been banned for some time under long-standing legislation.

Problems of enforcement

The environmental legislation described in this chapter should ensure that the environment is protected from most forms of pollution, barring unavoidable accidents. However this is not always the case, for two reasons. Firstly, several of the Acts dealing with pollution have been 'enabling' acts under which regulations must be made before the aims of the law can be achieved. Such regulations have not always appeared with any speed and in some instances appropriate measures have not been taken at all.

Secondly, if laws are to work they must be enforced. There are problems of recruitment in many agencies responsible for controlling pollution and, even when fully staffed, some are simply too small to deal with the problems arising. HMIP has national responsibilities yet has suffered from understaffing for years, despite increased workloads. Environmental Health Officers have an increasingly important role in pollution control at a local level, yet they are short of staff in many places and have even been cut back in some cases. As a result, complaints may not be dealt with adequately and preventive work and monitoring may be halted. Even when cases are brought to court, penalties can be derisory, although the Environmental Protection Act has increased penalties for some offences and a £1 million fine on Shell for polluting the Mersey in 1989 had a salutary effect. There is a distinct possibility that the directors of offending companies may receive prison sentences for some offences in the future, which may act as a powerful deterrent.

5: WHAT YOU CAN DO

There are two ways in which you can help solve pollution problems. One is to persuade others to change their policies and practices and the other is to change your own. Increasingly, people are choosing to do both.

In order to change the practices of other people – especially politicians – it is often most effective to join a pressure group. There are several national and international organisations dealing with pollution (see pages 139–141) and these welcome new members, especially those who can contribute time as well as money. They have groups in many towns and villages but if there is not one near you then it is usually easy to form one, with help and support from the organisation's headquarters.

You may wish to campaign on local issues such as a polluted river, rubbish dumping or a smoky factory. Or, you may wish to tackle global issues such as the greenhouse effect by promoting energy conservation, or saving the ozone layer by persuading DIY stores and building suppliers to stock CFC-free products. A local campaigning group is an effective way to do this and it is usually possible to obtain useful publicity in local media.

You need not join a group to have a voice if you do not want to. An individual writing clear and coherent letters to a newspaper or pressurising councillors and Members of Parliament can be an influential force, particularly if there are thousands of such people around the country. One of Friends of the Earth's early successes, a ban on the import of some whale products, was acknowledged to be the result of a letter-writing campaign to the responsible minister, backed up by technical material from the group's head office. Ministers and Members of Parliament receive dozens of circulars and a fair amount of abusive correspondence; temperate individual letters on specific issues tend to be taken more seriously. You could consider raising a petition in support of an anti-pollution measure, although you will need a large number of signatures to have any effect on national policies. Nevertheless, doing so can attract worthwhile publicity for the campaign and also makes the people you approach think about the issue.

Politicians make the laws but they are enforced by a wide range of bodies as described in the previous chapter. They may be susceptible to public pressure and some will positively welcome help from people in carrying out their duties. The National Rivers Authority cannot have permanent patrols on every river, so if you see an oil slick or dead fish floating on a river, tell them about it. Environmental Health Departments and Waste Regulation Authorities will also be pleased to hear of possible breaches of pollution laws, be it chimneys emitting excessive smoke or suspicious drums being dumped off the back of a lorry into a field.

When it comes to changing your own behaviour in order to reduce pollution, there are many things that you can do. Several books have been produced detailing how you can 'go green'; some are listed in the bibliography. Some of their points, and some new ones, are worth repeating here.

Everything we do or buy has some implication for the environment. Any use of energy carries with it a pollution load, be it carbon dioxide released to the atmosphere or radioactivity generated for subsequent disposal. The exception to this is that some renewable energy sources generate little pollution in use, although they are not without other environmental impacts. Every item we buy uses materials which involved the generation of pollution during extraction, processing and transport; and even if the item is not polluting in use, its disposal could generate pollutants.

The first step in personal pollution control is therefore to use less. This does not mean an instant switch to austerity or a total abandonment of all the benefits of twentieth-century living. What it does entail is an avoidance of waste in all its forms – such as the purchase of disposable items where a reusable one will do, the use of excessive packaging, or the inefficient use of energy. Of course, this means going against the ethos of a consumer society and this takes some willpower – advertisers are very good at convincing us that we need all sorts of products which are far from essential. This seems to be getting worse as mail order companies bombard households with catalogues full of trivia.

In addition to using less, we can reduce the creation of pollution by choosing carefully what we buy, since not all products pollute equally. To give a simple example, it is possible to buy many products in either a pump spray or an aerosol can. The former is refillable, simple in construction and the material from which it is made may eventually be recyclable (although it is not usually recycled at the moment). An aerosol can, however, is complex, which means more energy in manufacture; uses a mixture of materials which prevent recycling; cannot be reused; and if it does not emit CFCs may well release greenhouse gases in use. Green consumerism in itself will not solve our pollution problems but it can alleviate some of them – and it will send signals to manufacturers and retailers, telling them that the public is concerned about environmental issues.

The end of the chain, waste disposal, is a final point at which we can act to reduce pollution. Composting organic wastes may involves the release of carbon dioxide from the compost heap but this is less damaging than the uncontained release of methane from a landfill, since methane is a more powerful greenhouse gas. Gardeners who use compost are also avoiding the pollution created by fertiliser manufacture. Recycling wastes is normally cited as a means of conserving natural resources, but when the resource in question is renewable the main benefits come from reduction in pollution and energy use – savings of 50% are possible in both these areas when recycled paper is made compared with paper made from fresh pulp. Recycling waste motor oil is particularly useful, since oil poured down drains can cause serious pollution, and even taking glass to a bottle bank saves energy and hence pollution – unless you drive a long distance just to deposit a few bottles.

The list of points on how to reduce personal pollution which follows is not completely comprehensive but it will provide you with some ideas. No doubt you can think of others. They are divided into particular areas of activity but there is obviously some overlap between categories.

Energy conservation

Energy can be saved in many ways in the home, from insulating roofs and walls to lagging hot water tanks and pipes. Double glazing may help you to feel warmer by cutting down draughts but other measures tend to be more cost effective. Simple draught-proofing measures around doors, letterboxes and windows are very effective, but be sure to leave sufficient ventilation for gas appliances. If your house is affected by radon (q.v.) you will need better ventilation: consult your local Environment Health Department for advice. Your choice of fuel is important. Electricity is very inefficient as a means of heating homes, since two-thirds of the energy in the fuel is thrown away at the power station. Gas is much more efficient, especially if burned in modern condensing boilers. Open coal fires are inefficient – enclosed solid fuel boilers are much better – and you should always use a smokeless fuel. Making burnable bricks out of old newspapers is laborious and burning them may be illegal in smoke control areas; it is far better to recycle the papers. Wood stoves can be highly polluting.

One question you should ask yourself is 'How warm does the house have to be?' It is environmentally sounder and also cheaper to put on a jumper than to turn up the central heating during a cold spell, although elderly people and those who are ill may need warmer temperatures. Individual radiator thermostats provide effective control over the temperature in each room and mean that you do not need to have a bedroom as warm as a living room. Rooms in which babies sleep should not be too hot – overheating has been implicated in cot deaths. Separate thermostats on hot water tanks can also save energy. Computerised control systems for central heating boilers are now quite cheap.

Electrical appliances are often wasteful – insulation levels in freezers and refrigerators are often poor, so you should check on a product's energy consumption before you buy it as well as checking on its CFC content. The location of such items in the kitchen is also important – putting a fridge next to a cooker means that it has to work much harder to keep cool. Microwave ovens use less energy than conventional cookers, especially for dishes such as baked potatoes, but they should be used carefully for food safety reasons. Energy-saving light bulbs are now becoming widely available – they save money in the long run although they are expensive to buy.

Water usage

Economising on water usage may seem to have little to do with preventing pollution but most water is pumped some distance or raised to a height to ensure sufficient pressure; hence reducing water use saves energy. Hot water has obviously been heated with fuel and the link is more direct. There are several ways of economising on water in the household, from ensuring that taps do not leak to using washing-up water on the garden. Water can be collected in a water butt and used for the garden or washing windows or the car – if you have one – but this is not recommended for cooking or drinking. Dual-flush toilet cisterns can save considerable amounts of water, as can putting a plastic bag full of water in an existing cistern.

Transport

The way in which we move about has serious implications for pollution. Even the most carefully engineered motor car uses energy, emits carbon dioxide and also releases other pollutants. Its manufacture also involves the production of pollution as do the raw materials from which it is made. As a society we cannot avoid using cars, especially in rural areas, but as individuals we can reduce our usage of private motor vehicles to a considerable degree. Reasonably full buses and trains are more efficient in terms of fuel consumption per passenger-mile than private cars and the amounts of pollutants emitted per passenger-mile are smaller. Bicycles and feet are even more efficient for short journeys and even less polluting – and they can be healthier for you.

If you do need to use a car, consider hiring one if you only need it occasionally. Once you own a car the temptation to use it unnecessarily is great. Choose one that runs on unleaded petrol and is also fitted with a catalytic converter – and make sure that it returns a good mileage per gallon. Drive at moderate speeds, without over-revving the engine, and smoothly; accelerating rapidly burns up large amounts of fuel and produces much more pollution than gradual acceleration. High speed driving uses much more fuel for the same distance travelled than does slower driving. Plan your journeys so that you can complete several tasks in one go – e.g. taking children to school, collecting shopping and visiting a relative – rather than making three expeditions. If you need to travel to work by car, try to get involved in a car sharing scheme.

Household products

Many household products, such as cleaning materials, are now being marketed as 'green' or 'environment-friendly' and it can be very difficult to choose the best option. As far as washing powders are concerned, all are now at least 90% biodegradable. Where some brands offer environmental benefits over others is in the ingredients other than the detergent chemicals which they contain. Some products are marketed as phosphate-free and in areas such as Norfolk this can be beneficial in helping to prevent eutrophication. However, phosphate manufacture usually involves the release of cadmium to the marine environment, so buying a phosphate-free detergent wherever you live can be beneficial. The environmental effects of some bleaches and optical brighteners are uncertain but it makes sense to avoid them as they are not generally necessary. Separate washing bleaches can be bought if they are needed for especially dirty clothes. Concentrated products may offer advantages over more dilute versions since they involve less packaging and transport, although the statement that a product contains 'X% less chemicals' is nonsense.

Misleading claims have been made about other products as well. Phosphates have never been added to washing-up liquid, for instance; and why one manufacturer is advertising that its polish is nitrate-free is a mystery, since nitrates would not normally be added to such products. It is a bit like advertising that Hyde Park is free of rhinos.

Some polishes and cleaners contain solvents such as white spirit and these should be avoided if possible (which may be difficult as they are not always labelled) since they may contribute to the greenhouse effect. Some chlorinated solvents may also damage the ozone layer. The use of chlorine-based disinfectant bleaches may lead to the formation of dioxins in the environment and alternatives, containing hydrogen peroxide, are now available.

Batteries for electrical appliances can cause pollution when they are disposed of, because of their heavy metal content. Mercury is often present and rechargeables contain nickel and cadmium. If you can, use mains-operated equipment rather than batteries because any battery is inefficient in energy use when compared with mains power. Transformers which provide the low voltage needed by some appliances are available from electrical shops and many radios and cassette players can run directly from the mains. If you need to use a battery, try to use a rechargeable version and attempt to recycle it when it no longer works; at least one manufacturer is operating a recycling scheme in the UK and others may follow. If you cannot use rechargeables (which save money as well as reducing pollution) then pick a brand which is mercury free.

Do-it-yourself

Many home maintenance and decoration activities can cause pollution as a wide range of harmful materials has been used in the past and some are still in use. One of the most serious is lead which was a major component (up to 50%) of some paints before 1939 and was still used, in smaller quantities, in the 1980s. This means that anyone removing old paint from surfaces should assume that it contains lead. Indeed, it may be better to leave old paint in place if it is in sufficiently good condition for overpainting since any removal process risks releasing the toxic metal into the environment and may cause lead poisoning, a common ailment during urban lead paint removal programmes in the US in the 1970s.

If old paint has to be removed, it should *never* be rubbed down with dry sandpaper or abrasive attachments to power tools, since this generates a fine dust which is easily inhaled or picked up by children. Hot air strippers can be used providing they do not generate temperatures higher than 500 degrees C, but blowlamps should not, as their high temperatures can vaporise the lead. Chemical strippers can be dangerous and some solvents used are greenhouse gases. Any paint residues produced should be carefully wrapped and placed in the dustbin and the area involved thoroughly cleaned before children or food are allowed in the room. Residual paint on the stripped surfaces should be rubbed off with wet sandpaper and the waste disposed of carefully.

New paint may release solvents as it dries and these can contribute to air pollution. Low solvent paints are becoming more readily available and should be used where possible. White spirit and other paintbrush cleaners should be used in the minimum quantities possible.

Wood preservatives constitute another class of potentially harmful materials and the books *C for Chemicals* and *Toxic Treatments* provide more details (see Bibliography). In general it is more effective to prevent conditions favouring rot from occurring – e.g. by stopping leaks and damp – than to rely on chemicals to control the problem. Lindane should not be used as a woodworm killer. If you have bats in your roof, you should take professional advice before using *any* chemicals there, as bats are protected by law.

Other household pest control products may be hazardous – dichlorvos, for instance, is used in fly strips and some veterinary products, yet it is toxic and suspected of causing cancer. Flies can sometimes be trapped on flypapers and the occasional fly or wasp is better swatted than poisoned by a spray or strip which fills the whole room with vapour. Garden pests can be controlled, or at least reduced, by a number of measures which do not involve the use of seriously harmful chemicals; details are available from organisations listed in Chapter 6 who can also supply information about gardening without artificial fertilisers. Do not assume, however, that because a substance is natural it is harmless. Derris, a natural insecticide, was originally used as a fish poison while nicotine is even more toxic.

Food

Food production, in many cases, is a polluting business – indeed some types of farming resemble an industrial process more than a rural activity. Intensive livestock units produce highly polluting concentrated slurry and emit ammonia, while arable farming can lead to fertiliser runoff into rivers, damage to wildlife from insecticide use and groundwater pollution by nitrates and pesticides. Thus one simple means of reducing the pollution impact of food is to choose organically grown produce. This is available in many supermarkets now and, although it often costs more, may become less expensive as demand increases. If you eat meat, free-range organic meat and eggs are likely to involve the generation of less pollution than intensively produced equivalents.

Food processing is an energy-intensive activity that can lead to river pollution when high-BOD effluents are generated. Although there may be economies of scale in factories it is generally more efficient and less wasteful to buy food in a relatively unprocessed state and prepare it yourself, rather than buying ready-prepared meals.

Packaging

Many of the goods we buy are wrapped in more packaging than is necessary to protect the product from damage or prevent leakage. This may be because the wrapping is used as a selling point or because certain types of pack are easier to display or sell in a supermarket. There is obviously a pollution implication in excess packaging since energy and raw materials are needed to make it and it may cause problems when disposed of. There are many examples of excessive packaging – a battery which is wrapped in plastic, shrink-wrapped onto a card and then put in a plastic bag when sold; apples shrink-wrapped onto a plastic tray and put in a plastic bag; most products in aerosol cans, and so on.

Some of this packaging is difficult to avoid since all manufacturers use the same techniques; but in some instances you can choose alternatives with minimal packaging such as loose fruit, nails and screws bought loose rather than in packages, and drinks in returnable bottles (although these are hard to find). Some packaging is recyclable or made from recycled materials – 'green' household products are often packaged in this way. You do not need to accept a plastic bag for goods in supermarkets – once you have paid for something it is yours and no-one has the right to take it from you and put it in a bag. As long as you have a receipt you cannot be accused of stealing it.

Some waste packaging can be used for other purposes; schools and playgroups often have uses for items such as yogurt pots, washing-up liquid bottles and cardboard tubes. This prevents or delays any problems which may be caused by disposing of the items but there is a limit to the number of things that can be used in this way. Beekeepers may welcome jam and honey jars and home brewers and winemakers can use empty bottles.

Recycling

Recycling, as opposed to re-use, means reclaiming raw materials from waste and this is currently carried out (on domestic waste) with paper, textiles, glass, oil, aluminium and, in some cases, certain plastics. Recycling saves energy – making aluminium from scrap rather than fresh ore takes about a thirtieth of the energy – and also reduces pollution from the manufacture of the material. It also reduces pollution from waste disposal.

Friends of the Earth have published local recycling guides which are widely available and tell you where to take waste materials. Unfortunately recycling, especially of paper, is subject to fluctuations in the market and there is often a glut of paper. In 1989 some newspaper collected for recycling had to be burnt or dumped as it could not be sold. Using products made from recycled materials helps to maintain a demand, especially if the product is made from comparatively low-grade waste such as old newspapers.

Disposability

The lifetime of a product greatly affects its polluting potential since an appliance which has to be replaced annually is likely to have a much greater environmental impact than one which lasts five years and has comparable pollution 'running costs'. Although reliability and long lifetime is prompted as a feature in some instances – e.g. certain types of car – many products are not designed to have very long lifetimes, not least because manufacturers would like to sell you a replacement before too long. Fashions in clothes, furnishings and other accessories also demand regular replacement with consequent increases in energy use and pollution. Other items are deliberately marketed as being disposable – razors, cutlery, plates and even underwear are obvious examples – and these inevitably result in more pollution than their reusable counterparts.

In general, to reduce pollution it is better to buy products which are well-made and reliable, which can be re-used and will last a long time. A further bonus would be that they could be recycled or used for something else once they are finished with. Secondhand items can be good buys in this respect – even fairly ordinary furniture made fifty years ago is quite serviceable today, whereas modern chipboard-and-staples furniture often disintegrates in ten years or less, never mind fifty.

The materials from which an item is made are relevant in this respect since some are inherently more durable than others. Furniture and carpets made from natural materials may be more resistant to damage – e.g. by heat – than their counterparts made from synthetic materials. Many plastics become brittle over the years. Solid wood is also easier to repair than chipboard and well-made wooden sash windows will outlast modern plastic window frames with built-in obsolescence. Car exhausts made from stainless or aluminised steel will last much longer than mild steel systems and galvanised steel is more durable than painted metal.

When things do begin to wear out or become damaged repairability is important. Anything which delays the replacement of an item is desirable as long as it does not involve an inordinate amount of energy or materials use. Unfortunately the tendency with many manufacturers is to scrap a large part of a defective appliance rather than attempt to repair it since it saves on labour charges.

The need for commitment

Changing one's own lifestyle by becoming greener is obviously worthwhile, but it is even more effective if others can be persuaded to do the same. This can be done through pressure groups, as described above, but it can also be achieved through education of individuals. To do this you need to be well-informed about environmental problems and how to alleviate them. If you are, you can begin to take the green message to your family, to people at work, your friends and anyone else who is prepared to listen! If you are a member of any organisation, from a church to a school's board of governors, you can try to persuade that group to use less polluting products, be they cleaning materials or recycled paper. You can promote energy conservation – it is, after all, financially beneficial and also accords with stated government policy. You can even organise car-sharing schemes for members travelling to meetings or set up a collection of waste paper or aluminium.

The list of possible actions which the individual can take, alone or in groups, is endless and this chapter can only list a few of them. One thing is clear; pollution problems are never going to be solved if we wait for governments, national organisations and companies to do it on their own initiative. They need encouragement, praise where it is due and criticism where necessary. But as well as exerting pressure a personal commitment to reducing pollution is necessary, a commitment which ranges from limiting our use of energy to deciding which brand of detergent to buy. Going green may not necessarily be the easy option but it may be the only one: we cannot all be Secretary of State for the Environment, but we can all make that personal commitment.

6: USEFUL ORGANISATIONS

There are many organisations concerned with the environment and pollution, ranging from official government bodies to local pressure groups. Local organisations can usually be found by contacting reference libraries which will also have details of branches of national bodies. The following organisations can all be contacted for information although the stance taken on a particular issue will depend on the attitude and purpose of the organisation in question. When contacting pressure groups, please remember to enclose a stamped addressed envelope as they are usually short of funds.

Pressure groups and similar organisations

Ark Campaigns, *498 Harrow Road, London W9 3QA (Tel. 081 968 6780).* Ark has a growing number of local groups, which campaign on many environmental issues including pollution, and a national campaign strategy which includes combatting the greenhouse effect. They raise funds by, among other things, marketing consumer products designed to have a reduced impact on the environment when compared with conventional versions.

CLEAR, *3, Endsleigh Street, London WC1H 0DD (Tel. 071 278 9686).* CLEAR, the Campaign for Lead-free Air, was set up to campaign for unleaded petrol and in this it was successful. It is still concerned about other sources of lead in the environment.

Foresight, *The Old Vicarage, Church Lane, Witley, Surrey GU8 5PN (Tel. 042868 4500).* An information organisation concerned about the effects of nutrition and chemical contamination, before and after conception, on unborn children and their parents. It publishes a number of advice leaflets.

Friends of the Earth, *26-28 Underwood Street, London N1 7JQ (Tel. 071 490 1555).* FOE is one of Britain's foremost pressure groups and is the sister organisation of many other FOE groups worldwide. It campaigns on many environmental issues; on the pollution front it has tackled pesticides, river quality, drinking water, vehicle emissions, the ozone layer, the greenhouse effect and waste dumping. It has an extensive network of local groups and a large central staff.

Greenpeace, *30-31 Islington Green, London N1 8XE (tel. 071 354 5100).* Also part of an international organisation. It addresses issues which have a global dimension, particularly in the marine sphere, and has campaigned on the dumping and incineration of wastes at sea, acid rain and coastal pollution as well as several other pollution topics.

Henry Doubleday Research Association, *Ryton-on-Dunsmore, Coventry CV8 3LG (Tel. 0203 303517).* The HDRA promotes organic gardening by carrying out research, providing information and marketing seeds, environmentally benign pest control methods and fertilisers. They publish a catalogue of their products as well as a number of booklets.

The Food Commission, *88 Old Street, London EC1V 9AR (Tel 071 253 9513).* An independent body concerned about the safety of food. It carries out research, publishes books and reports, and campaigns for safer food. It has tackled the issues of pesticide residues and other contaminants in food. Formerly the London Food Commission.

London Hazards Centre, *3rd Floor, Headland House, 308 Grays Inn Road, London WC1X 8DS (Tel. 071 837 5605).* Researches and campaigns on health and safety issues affecting people at work, at home and in the community. These include exposure to chemicals such as pesticides and asbestos. The group publishes a bulletin, *The Daily Hazard,* five times a year.

The Marine Conservation Society, *9 Gloucester Road, Ross-on-Wye, Herefordshire HR9 5BU (Tel. 0989 66017).* This organisation is concerned with, among other things, pollution of the sea and publishes a list of clean beaches where bathers can swim in water free of raw sewage.

National Society for Clean Air and Environmental Protection *136 North Street, Brighton BN1 1RG (Tel. 0273 26313).* One of Britain's oldest pressure groups, the NSCA is concerned about all aspects of air pollution and noise. It has recently extended its brief by the addition of the 'environmental protection' phrase as it is concerned with integrated pollution control. It publishes the respected journal, *Clean Air,* and has a comprehensive reference library on air pollution.

Parents for Safe Food, *Britannia House, 1-11 Glenthorne Road, London*

W6 0LF (Tel. 081 748 9898). A small but effective pressure group campaigning against pesticide residues and other contaminants of foodstuffs. They have leaflets available on various contaminants.

Pesticides Exposure Group of Sufferers (PEGS) *10 Parker Street, Cambridge CB1 1JL (Tel. 0223 64707).* A group of people who have suffered from the effects of pesticide exposure and who are pressing for tighter controls on these substances.

Pesticides Trust, *c/o WUS, 20 Compton Terrace, London N1 2UN (Tel. 071 354 3860).* An independent research and campaigning organisation concerned about the hazards of pesticides and the promotion of alternatives.

Soil Association, *86-88 Colston Street, Bristol BS1 5BB (Tel. 0272 290661).* The Soil Association is concerned with organic farming and gardening and has a wealth of information on alternatives to pesticides and inorganic fertilisers.

Womens Environmental Network, *287 City Road, London EC1V 1LA (Tel. 071 490 2511).* A campaigning and information group dealing with environmental issues of particular interest to women, such as the effects of sanitary protection products and disposable nappies.

Official Bodies

The Department of the Environment, *2, Marsham Street, London SW1P 2EB (Tel. 071 276 3000).* The DOE is the main government department dealing with pollution, although its policy making is often hamstrung by other departments such as the Department of Transport and the Treasury. It can provide information on current government policy and also publishes some useful reports on pollution issues.

The Department of Trade and Industry, *Warren Spring Laboratory, Gunnels Wood Road, Stevenage, Herts. SG1 2BX (Tel. 0438 741122).* The Warren Spring Laboratory carries out original research on air pollution and waste management issues, the latter topic centring around recycling. It provides advice to businesses as part of the DTI's Environmental Programme and also publishes reports for general circulation.

The Drinking Water Inspectorate, *2, Marsham Street, London SW1P 2EB (Tel. 071 276 3000).* This body is responsible for maintaining standards for drinking water after the privatisation of the water industry.

Environmental Health Departments of local authorities (district councils in most areas: borough councils in London) are responsible for monitoring some types of air pollution and enforcing the law – see Chapter 4. They may also monitor drinking water and, where appropriate, bathing beaches.

Harwell Laboratory, *Harwell, Didcot, Oxfordshire OX11 0RA (Tel. 0235 24141).* The Harwell Laboratory of the United Kingdom Atomic Energy Authority is a centre of expertise on all aspects of waste management and the pollution generated therefrom. It publishes the monthly bulletin, *Waste Management Today,* and also provides information on action to be taken in chemical emergencies such as chemical tanker crashes.

The Health and Safety Executive has regional offices in major towns. It is responsible for monitoring health and safety in industry and enforcing the relevant legislation. It has considerable expertise in the field of chemicals, from the point of view of health effects and major hazards, and publishes some useful guides and reports available from HMSO.

Her Majesty's Inspectorate of Pollution, located in the Department of the Environment, is responsible for controlling major sources of air pollution and the management of hazardous waste. It will implement the philosophy of Integrated Pollution Control promoted by the Environmental Protection Act. It has regional offices as well as the national headquarters in Marsham Street.

National Radiological Protection Board, *Chilton, Didcot, Oxfordshire OX11 0RQ (Tel. 0235 831600).* The body which sets standards for, and advises on, radiation protection in the UK.

National Rivers Authority, *30-34 Albert Embarkment, London SE1 7TL (Tel. 071 820 0101).* The NRA is the body which controls discharges to rivers, taking samples and bringing prosecutions where appropriate. It has a number of regional offices and is moving its headquarters to Bristol in 1991.

Royal Commission on Environmental Pollution, *Church House, Great Smith Street, London SW1P 3BL (Tel. 071 276 2080).* An independent body which regularly enquires into pollution issues. Its reports are extremely useful and authoritative.

Water Research Centre, *Medmenham, Henley Road, Marlow, Bucks. SL7 2HD (Tel. 0491 571531).* A company set up from within the water industry, WRC is a centre of research and expertise on all aspects of the water cycle. It produces reports, some of which are confidential and some publicly available.

7: BIBLIOGRAPHY AND COURSES

There are many different books, magazines and journals dealing with the environment in general and pollution in particular. It is impossible to read them all, but the following is a selection of titles which may be useful for further reading.

Magazines and journals

BBC Wildlife   A general countryside and environment magazine which covers some pollution topics. Available in newsagents.

Clean Air   The quarterly journal of the National Society for Clean Air and Environmental Protection. *Clean Air* contains features on various aspects of air and noise pollution and updates on legislation and other developments.

Daily Hazard   A bulletin produced by the London Hazards Centre covering various types of work and community hazards.

The Ecologist   All types of environmental issues, from the technical to the philosophical, are covered in *The Ecologist,* Britain's longest-established environmental magazine. The subscriptions department is at Worthyvale Manor, Camelford, Cornwall PL32 9TT.

ENDS Report   An invaluable monthly report covering new technical and legislative developments in the environmental sphere as well as profiles of companies' environmental activities and analyses of national and international policies. Details from ENDS at Finsbury Business Centre, 40 Bowling Green Lane, London EC1R 0NE.

Green Magazine   A wide-ranging environmental monthly magazine carrying topical articles and lavish illustrations. Available in newsagents.

New Scientist   A weekly science magazine which gives thorough coverage to many environmental issues in both features and news sections. Available in newsagents.

Warmer Bulletin   A free magazine covering the recycling and use of waste as fuel, with coverage of other aspects of energy conservation and pollution prevention. Details from the Warmer Campaign, 83 Mount Ephraim, Tunbridge Wells, Kent TN4 8BS.

Books

Acid Rain, Fred Pearce (Penguin, 1987). A useful account of the acid deposition problem as understood in 1987.

Being Green Begins at Home, Mike Birkin et al. (Green Print, 1990). A practical and positive guide to an environmentally-friendly lifestyle.

Bitter Harvest, Joyce Egginton (Martin Secker and Warburg, 1980). A chilling account of how the whole food chain in Michigan was contaminated by the accidental mixing of a toxic chemical into animal feed.

C for Chemicals, Mike Birkin and Brian Price (Green Print, 1989). A guide to the hazards of household and garden chemicals and how to avoid them.

Dictionary of the Environment, Steve Elsworth (Paladin, 1990). A wide-ranging and valuable guide to key environmental issues, packed with useful statistics.

Dictionary of Environmental Science and Technology, Andrew Porteous (Open University Press, 1991). A comprehensive and authoritative guide to the science and technology of the environment.

EEC Environmental Policy and Britain, Nigel Haigh (Longman). Updated regularly, this book gives an authoritative account of EC environmental policy and how Britain implements it (or not).

Ecotoxicology, F. Moriarty (2nd Edition, Academic Press, 1988). An academic book dealing with the effects of pollutants on ecosystems.

The Energy Alternative, Walter Patterson (Boxtree, 1990). A thorough and extremely readable account of how our use of energy must change if we are to avoid wrecking the planet.

Food Adulteration, London Food Commission (Unwin, 1988). A comprehensive account of the materials added to food, deliberately or accidentally, including nitrates, pesticides and food poisoning organisms.

The Green Consumer Guide, John Elkington and Julia Hailes (Gollancz, 1988). A practical guide to choosing less environmentally damaging products and services. A 'supermarket' version is also available.

How Safe is Safe? Barrie Lambert (Unwin Paperbacks, 1990). A thorough discussion of current radiation controversies from safety at Sellafield to medical X-rays.

How to be Green, John Button (Century, 1989). A lifestyle guide on how to reduce personal impacts on the environment, produced with the collaboration of Friends of the Earth.

Introduction to Toxicology, J.A. Timbrell (Taylor and Francis, 1989). An academic introduction to how poisons of all types work.

The Lead Scandal, Des Wilson (Heinemann Educational, 1983). An account, by one of the leading participants, of the campaign to remove lead from petrol.

Lead versus Health, edited by Michael Rutter and Robin Russell Jones (John Wiley, 1983). A series of papers on the sources and effects of low-level lead exposure.

Multiple Exposure, Catherine Caufield (Martin Secker and Warburg, 1989). A frightening account of the misuse of radiation, from its discovery in the last century to contemporary problems with nuclear power and weapons.

Nitrates, Nigel Dudley (Green Print, 1990). An account of all aspects of the nitrate fertiliser problem including alternative methods of crop husbandry.

Pesticides, Chemicals and Health, British Medical Association (Edward Arnold, 1991). An authoritative report, brought out in a book form, of the hazards of pesticides and some other chemicals.

Pesticides and Your Food, Andrew Watterson (Green Print, 1991). A layperson's guide to the hazards of pesticides and their residues in the food we eat.

Pesticides Health and Safety Handbook, Andrew Watterson (Gower Technical, 1988). A directory of the main pesticides in use and their hazards to exposed people.

NSCA Pollution Handbook. Produced annually by the National Society for Clean Air and Environmental Protection, this handbook is a comprehensive guide to pollution legislation and also contains useful material on the effects of pollution and its control.

The Poisoned Womb, John Elkington (Viking, 1985). A disturbing account of how all aspects of human reproduction may be impaired by pollutants.

Poisoners of the Sea, K.A. Gourlay (Zed Books, 1988). An account of all aspects of marine pollution, especially the dumping of wastes from ships.

Silent Spring, Rachel Carson (Penguin, 1965). The original devastating account of environmental damage by pesticides. The debate has moved on since it was published, but the principles remain the same.

Toxic Treatments, London Hazards Centre (London Hazards Centre, 1988). An account of the risks posed by wood preservatives, at work and at home, to people exposed to them.

Turning up the Heat, Fred Pearce (Bodley Head, 1989). A very readable treatment of the greenhouse problem and related effects of pollutants on climate.

The above list cannot be completely comprehensive as there are many more books available dealing with pollution and the environment. Organisations such as Friends of the Earth frequently publish books and reports on pollution topics, while the Royal Commission on Environmental Pollution produces extremely useful reports from time to time.

Courses

The Open University runs two courses which cover pollution matters for people who are not necessarily studying for a degree. The course 'Environment' looks at a range of environmental issues and requires no scientific background, while 'Environmental Control and Public Health' requires some knowledge of chemistry and a little mathematics. Details from ASCO, The Open University, PO Box 76, Milton Keynes, MK7 6AN.

INDEX

Index of pollutants and processes (covering Parts 1 and 2). Main entries in Part 2 appear in bold type.